The Truth About
COVID-19

Also by Dr. Joseph Mercola

*EMF*D: 5G, Wi-Fi and Cell Phones: Hidden Harms and
  How to Protect Yourself* (2020)

*KetoFast: Rejuvenate Your Health With a Step-by-Step Guide
  to Timing Your Ketogenic Meals* (2019)

*Fat for Fuel: A Revolutionary Diet to Combat Cancer,
  Boost Brain Power, and Increase Your Energy* (2017)

*Effortless Healing: 9 Simple Ways to Sidestep Illness, Shed Excess
  Weight, and Help Your Body Fix Itself* (2015)

*The Great Bird Flu Hoax: The Truth They Don't Want You to Know
  About the 'Next Big Pandemic'* (2006, with Pam Killeen)

*The No-Grain Diet: Conquer Carbohydrate Addiction and Stay Slim
  for Life* (2003, with Alison Rose Levy)

*Dark Deception: Discover the Truth About the Benefits of Sunlight
  Exposure* (2008, with Jeffry Herman)

*Generation XL: Raising Healthy, Intelligent Kids in a High-Tech,
  Junk-Food World* (2007, with Ben Lerner)

Healthy Recipes for Your Nutritional Type (2007, with Kendra
  Degen Pearsall)

Take Control of Your Health (2007, with Kendra Degen Pearsall)

*Sweet Deception: Why Splenda®, NutraSweet®, and the FDA May Be
  Hazardous to Your Health* (2006, with Kendra Degen Pearsall)

Also by Ronnie Cummins

*Grassroots Rising: A Call to Action on Climate, Farming, Food,
  and a Green New Deal*

Genetically Engineered Food a Self-Defense Guide for Consumers
  (2000, with Ben Lilliston)

Children of the World (1990, a five-book series)

The Truth About

COVID-19

Exposing The Great Reset, Lockdowns, Vaccine Passports, and the New Normal

---

Dr. Joseph Mercola
Ronnie Cummins

Foreword by Robert F. Kennedy Jr.

Chelsea Green Publishing
White River Junction, Vermont
London, UK

Figure 3.1, on page 51, courtesy of Mercola.com.
Figure 6.1, on page 92, courtesy of Cronometer.com.
Figure 6.2, on page 93, courtesy of Cronometer.com.
Figure 7.1, on page 119, courtesy of Mercola.com.

Project Manager: Patricia Stone
Project Editor: Brianne Goodspeed
Proofreader: Nancy Crompton
Indexer: Linda Hallinger
Designer: Melissa Jacobson

Printed in the United States of America.
First printing April 2021.
10 9 8 7 6 5 4 3 2 1   21 22 23 24 25

ISBN 978-1-64502-088-2 (hardcover) | ISBN 978-1-64502-089-9 (ebook)
| ISBN 978-1-64502-090-5 (audio book)

Library of Congress Control Number: 2021933705

Chelsea Green Publishing
85 North Main Street, Suite 120
White River Junction, Vermont USA

Somerset House
London, UK

www.chelseagreen.com

To the rebels, dreamers, and visionaries
who will regenerate the Earth

Contents

Foreword

Government technocrats, billionaire oligarchs, Big Pharma, Big Data, Big Media, the high-finance robber barons, and the military industrial intelligence apparatus love pandemics for the same reasons they love wars and terrorist attacks. Catastrophic crises create opportunities of convenience to increase both power and wealth. In her seminal book, *The Shock Doctrine: The Rise of Disaster Capitalism,* Naomi Klein chronicles how authoritarian demagogues, large corporations, and wealthy plutocrats use mass disruptions to shift wealth upwards, obliterate the middle classes, abolish civil rights, privatize the commons, and expand authoritarian controls.

A consummate insider, the former White House Chief of Staff Rahm Emmanuel is known for his admonition that vested power structures should "never let a serious crisis to go to waste." But this tread-worn strategy—to use crisis to inflame the public terror that paves the road to dictatorial power—has served as the central strategy of totalitarian systems for the millennia.

The methodology is, in fact, formulaic, as Hitler's Luftwaffe commander, Hermann Göring, explained during the Nazi war crimes trials at Nuremberg: "It is always a simple matter to drag the people along whether it is a democracy, a fascist dictatorship, or a parliament, or a communist dictatorship. Voice or no voice, the people can always be brought to the bidding of the leaders. That is easy. All you have to do is tell them they are being attacked, and denounce the pacifists for lack of patriotism, and exposing the country to greater danger. It works the same in any country."

The Nazis pointed to the threats from Jews and Gypsies to justify the homicidal authoritarianism in the Third Reich. The dictatorial demagogue, Senator Joseph McCarthy, and the HUAC committee warned against communist infiltration of the State Department and film industry to rationalize loyalty oaths and the blacklist. Dick Cheney used the 911 attack to launch his "long war" against amorphous terrorism and the Patriot Act abridgments that laid the groundwork for the modern surveillance state. Now the medical cartel and its billionaire Big Tech accomplices have invoked the most potent, frightening, and enduring enemy of all—the microbe.

And who can blame them? Increasing the wealth and power of the oligarchy is seldom a potent vessel for populism. Citizens accustomed to voting for their governments are unlikely to support policies that make the rich richer, increase political and social control by corporations, diminish democracy, and reduce their civil rights. So demagogues must weaponize fear to justify their demands for blind obedience and to win public acquiescence for the demolition of civil and economic rights.

Of course, the first casualty must always be freedom of speech. After stoking sufficient panic against the hobgoblin *du jour*, robber barons need to silence protest against their wealth and power grabs.

In including free speech in the First Amendment of the Constitution, James Madison argued that all our other liberties depend on this right. Any government that can hide its mischief has license to commit atrocities.

As soon as they get hold of the levers of authority, tyrants impose Orwellian censorship and begin gaslighting dissenters. But ultimately they seek to abolish all forms of creative thinking and self-expression. They burn books, destroy art, kill writers, poets, and intellectuals, outlaw gatherings, and at their worst, force oppressed minorities to wear masks that atomize any sense of community or solidarity and prevent the subtle, eloquent nonverbal communication for which God and evolution have equipped humans with 42 facial muscles. The most savage Middle Eastern theocracies mandate masks for women, whose legal status—not coincidentally—is as chattels.

The free flow of information and self-expression are oxygen and sunlight for representative democracy, which functions best with policies annealed in the boiling cauldron of public debate. It is axiomatic that without free speech, democracy withers.

The most iconic and revered monuments of democracy therefore include the Athenian Agora and Speakers' Corner at Hyde Park. We can't help feeling exhilaration about our noble experiment in self-government when we witness the boisterous, irreverent debates in the House of Commons, or watch Jimmy Stewart's filibuster scene in *Mr. Smith Goes to Washington*—an enduring homage to the inseparable bond between debate and democracy.

To consolidate and fortify their power, dictatorships aim to replace those vital ingredients of self-rule—debate, self-expression, dissent, and skepticism—with rigid authoritarian orthodoxies that function as secular surrogates for religion. These orthodoxies perform to abolish critical thinking and regiment populations in blind, unquestioning obedience to undeserving authorities.

Instead of citing scientific studies to justify mandates for masks, lockdowns, and vaccines, our medical rulers cite WHO, CDC, FDA, and NIH—captive agencies that are groveling sock puppets to the industries they regulate. Multiple federal and international investigations have documented the financial entanglements with pharmaceutical companies that have made these regulators cesspools of corruption.

Iatrarchy—meaning government by physicians—is a little-known term, perhaps because historical experiments with it have been catastrophic. The medical profession has not proven itself an energetic defender of democratic institutions or civil rights. Virtually every doctor in Germany took lead roles in the Third Reich's project to eliminate mental defectives, homosexuals, handicapped citizens, and Jews. So many hundreds of German physicians participated in Hitler's worst atrocities—including managing mass murder and unspeakable experiments at the death camps—that the Allies had to stage separate "Medical Trials" at Nuremberg. Not a single prominent German doctor or medical association raised their voice in opposition to these projects.

So it's unsurprising that, instead of demanding blue-ribbon safety science and encouraging honest, open, and responsible debate on the science, the badly compromised and newly empowered government health officials charged with managing the COVID-19 pandemic response collaborated with mainstream and social media to shut down discussion on key public health and civil rights questions. They silenced and excommunicated heretics such as Dr. Mercola who refused to genuflect to Pharma and treat unquestioning faith in zero liability, shoddily tested, experimental vaccines as religious duty.

Our current iatrarchy's rubric of "scientific consensus" is the contemporary iteration of the Spanish Inquisition. It is a fabricated dogma constructed by this corrupt cast of physician technocrats and their media collaborators to legitimize their claims to dangerous new powers.

The high priests of the modern Inquisition are Big Pharma's network and cable news gasbags who preach rigid obedience to official diktats including lockdowns, social distancing, and the moral rectitude of donning masks despite the absence of peer-reviewed science that convincingly shows that masks prevent COVID-19 transmission. The need for this sort of proof is gratuitous.

They counsel us to, instead, "trust the experts." Such advice is both anti-democratic and anti-science. Science is dynamic. "Experts" frequently differ on scientific questions and their opinions can vary in accordance with and demands of politics, power, and financial self-interest. Nearly every lawsuit I have ever brought pitted highly credentialed experts from opposite sides against each other, with all of them swearing under oath to diametrically antithetical positions based on the same set of facts. Science is disagreement; the notion of scientific consensus is oxymoronic.

The modern intention of the totalitarian state is corporate kleptocracy—a construct that replaces democratic process with the arbitrary edicts of unelected technocrats. Invariably, their fiats invest multinational corporations with extraordinary power to monetize and control the most intimate parts of our lives, enrich billionaires, impoverish the masses, and manage dissent with relentless surveillance and obedience training.

In 2020, led by Bill Gates, Silicon Valley applauded from the sidelines as powerful medical charlatans—applying the most pessimistic projections from discredited modeling and easily manipulated PCR testing, and a menu of new protocols for coroners that appeared intended to inflate reporting of COVID-19 deaths—fanned pandemic panic and confined the world's population under house arrest.

The suspension of due process, due notice, and comment rulemaking meant that none of the government prelates who ordained the quarantine had to first publicly calculate whether destroying the global economy, disrupting food and medical supplies, and throwing a billion humans into dire poverty and food insecurity would kill more people than it would save.

In America, their quarantine predictably shattered the nation's once-booming economic engine, putting 58 million Americans out of work, and *permanently bankrupting over 100,000 small businesses, including 41,000 Black-owned businesses, some of which took three generations of investment to build.* These policies have also set into motion the inevitable dismantling of the social safety net that nurtured America's envied middle class. Government officials have already begun liquidating the 100-year legacies of the New Deal, New Frontier, the Great Society, and Obamacare to pay the accumulated quarantine debts. Say goodbye to school lunches, healthcare, WIC, Medicaid, Medicare, University scholarships, and more.

While obliterating the American middle class and dropping an additional 8 percent of Americans below the poverty line, the 2020 "COVID coup" transferred a trillion dollars of wealth to Big Technology, Big Data, Big Telecom, Big Finance, Big Media behemoths (Michael Bloomberg, Rupert Murdoch), and Silicon Valley Internet titans such as Jeff Bezos, Bill Gates, Mark Zuckerberg, Sergey Brin, Larry Page, and Jack Dorsey. It seems beyond coincidence that these men, who are cashing in on the poverty and misery caused by global lockdowns, are the same men whose companies actively censor critics of those policies.

The very Internet companies that snookered us all with the promise of democratizing communications have created a world where it has become impermissible to speak ill of official pronouncements, and practically a crime to criticize pharmaceutical products. The same Tech/Data and Telecom robber barons, now gorging themselves on the corpses of our obliterated middle class, are rapidly transforming America's once-proud democracy into a censorship and surveillance police state over which they profit at every turn.

For example, this cabal used the lockdown to accelerate construction of their 5G network of satellites, antennae, biometric facial recognition, and "track and trace" infrastructure that they, and their government and intelligence agency partners, will use to mine and monetize our data for free, compel obedience to arbitrary dictates, and to suppress dissent. Their government/industry collaboration will use this system to manage the rage when Americans finally wake up to the fact that this outlaw gang has stolen our democracy, our civil rights, our country, and way of life—while we huddled in orchestrated fear from a flu-like illness.

Predictably our other constitutional guarantees lined up behind free speech at the gibbet. The imposition censorship has masked this systematic demolition of our Constitution, including attacks on our freedoms of assembly (through social distancing and lockdown rules), on freedom of worship (including abolishing religious exemptions and closing churches, while liquor stores remain open as "essential service"), private property (the right to operate a business), due process (including the imposition of far reaching restrictions against freedom of movement, education, and association without rule making, public hearings, or economic and environmental impact statements), the Seventh Amendment right to jury trials (in cases of vaccine injuries caused by corporate negligence), our rights to privacy and against illegal searches and seizures (warrantless tracking and tracing), and our right to have governments that don't spy on us or retain our information for mischievous purposes.

Silencing Dr. Mercola's voice, of course, was the Medical Cabal's early priority. For decades, Dr. Mercola has been among the most effective and influential advocates against the pharmaceutical paradigm. He was an eloquent, charismatic, and knowledgeable critic of a corrupt system that has made Americans the world's top consumer of pharmaceutical drugs. Americans pay the highest prices for drugs, and have the worst health outcomes among the top 75 nations. Putting opiates—which kill 50,000 Americans annually—aside, pharmaceuticals are now the third biggest killer of Americans, after heart attacks and cancer.

Like a prophet in the wilderness, Dr. Mercola has argued for years that good health does not come in a syringe or a pill but from building strong immune systems. He preaches that nutrition and exercise are the most effective medicines, and that public health officials ought to be pushing policies that discourage reliance on pharmaceutical products and that safeguard our food supplies from Big Food, Big Chemical, and Big Ag. These predatory industries naturally consider Dr. Mercola to be Public Enemy #1.

Big Pharma's $9.6 billion annual advertising budget gives these unscrupulous companies control over our news and television outlets. Strong economic drivers (pharmaceutical companies are the biggest network advertisers) have long discouraged mainstream media outlets from criticizing vaccine manufacturers. In 2014, a network president, Roger Ailes, told me he would fire any of his news show hosts who allowed me to talk about vaccine safety on air. "Our news division," he explained, "gets up to 70% of ad revenues from pharma in non-election years."

Thus, pharmaceutical products were both the predicate and the punchline of the Cancel Culture. The Pharmedia long ago banned Dr. Mercola from the airwaves and newsprint while turning Wikipedia—which functions as Big Pharma's newsletter and propaganda vehicle—into a mill for defamations against him and every other integrative and functional health physician. At COVID's outset, the social media robber barons—all with their own financial entanglements with Big Pharma—joined the campaign to silence Mercola by ejecting him from their platforms.

It's a bad omen for democracy when citizens can no longer conduct civil, informed debates about critical policies that impact the vitality of our economy, public health, personal freedoms, and constitutional rights. Censorship is violence, and this systematic muzzling of debate—which proponents justify as a measure to curtail dangerous polarization—is actually fueling the polarization and extremism that the autocrats use to clamp down with evermore draconian controls.

We might recall, at this strange time in our history, my father's friend, Edward R. Murrow's warning that: "The right to dissent . . . is surely fundamental to the existence of a democratic society. That's the right that went first in every nation that stumbled down the trail to totalitarianism."

Robert F. Kennedy, Jr.

CHAPTER ONE

How the Pandemic Plans Unfolded

By Ronnie Cummins

*If it turned out COVID-19 came from a lab it would shatter the
scientific edifice top to bottom.*

—Tweet from Antonio Regalado, biomedicine editor
of *MIT Technology Review*[1]

COVID-19 and the profoundly misguided and self-destructive responses
to the pandemic have dragged us, kicking and screaming, or cowering
in fear, into the most serious world crisis since World War II. The pandemic
has shone a harsh spotlight on the fundamental *dis-ease* and ill health of
nations. The ongoing crisis has exposed our vulnerabilities to a variety of
life-threatening comorbidities, medical errors, and the corruption of the entire
pharmaceutical industry.

Beyond its effects on health and the health care industry, COVID-19 has
empowered the global elite more than ever before to manufacture lies and
half-truths. Uber-powerful Silicon Valley Big Tech corporations (Facebook,
Google, Microsoft, and Amazon), Big Pharma, the World Health Organization
(WHO), and philanthropic giant Bill Gates have indentured politicians and
scientists from across the political spectrum. The result is fearmongering, political polarization, and social engineering—all wrapped in a disguise of protection.

A shadowy network of military contractors and bioweapons specialists are
hiding behind the façade of biomedical and vaccine research while Big Tech
silences their critics. As Robert F. Kennedy, Jr., of Children's Health Defense
points out, the "Machiavellian manipulation of this pandemic amounts to
nothing less than an attempted coup d'état by big data, by big telecom, by big
tech, by the big oil and chemical companies and by the global public health
cartel led by Bill Gates and the WHO that . . . wants to magnify and amplify
its wealth and its power over our lives, over our liberties, that wants to subvert
our democracies and to destroy our sovereignty and our control over our lives
and our children's health."[2]

To understand and resolve this unprecedented crisis, we have no choice but to investigate, with a critical eye, the origins, nature, virulence, impacts, prevention, and treatment of COVID-19. We must examine both the "official story" of the pandemic—force-fed 24/7 to the public by the mass media, Big Tech, and the global public health establishment—and the genuine health threats posed by COVID-19, as a *highly transmissible biological trigger* that magnifies and intensifies preexisting chronic diseases and comorbidities. The elderly, as well as those with serious preexisting medical conditions such as obesity, diabetes, heart disease, lung disease, kidney disease, dementia, and hypertension, are underscoring the significant health changes that make us most vulnerable to COVID-19, as well as future pandemics.

Besides investigating the nature and the virulence of the SARS-CoV-2 virus itself, we must also examine the effectiveness and collateral damage of various government responses to the pandemic. This collateral damage includes the pandemic's impact on overall public health (mental and physical), increasing the number of deaths from chronic diseases going untreated as well as engendering a state of chronic stress among hundreds of millions of people.

We must also weigh the pandemic's impact on the economy, poverty, hunger, homelessness, and unemployment, as well as its role in increasing polarization and conflict in the body politic. And finally we must look at the alarming upsurge in authoritarian and totalitarian trends, including censorship, threats to privacy, restrictions on freedom of movement and assembly, health and consumer choice, local and regional sovereignty, and other basic human rights.

"Plandemic" or Anticipated Pandemic?

We will review the preponderance of evidence, which indicates that even though the SARS CoV-2 virus was lab-engineered, it was apparently accidentally, rather than deliberately, released. Although the COVID-19 pandemic was not necessarily *planned and then deliberately executed* on an exact time line by global elites, in fact this pandemic was *long anticipated*.[3]

The reckless "gain-of-function" experimentation—the scientific madness that weaponized SARS-CoV-2—was funded and carried out by partnership among the Chinese, American, and other governments, military, and Big Pharma, even after decades of lab accidents and dangerous releases of potential pandemic pathogens (PPPs) across the globe from scores of badly managed and relatively unregulated biomedical/bioweapons labs should have taught them better.[4]

This long-anticipated pandemic, officially announced by the World Health Organization on March 11, 2020, became utilized by an international network of powerful corporations and billionaires to dramatically expand their power, wealth, and control, in what can only be described as an attempted global coup d'état.

Since the coronavirus first emerged in Wuhan, China, in October–November 2019, SARS-Cov-2 (the virus) and COVID-19 (the disease attributed to the virus) have moved like a tsunami across the world, shocking and disrupting communities. In the COVID-19-driven time warp of the past 12 months, politics, economics, public opinion, and social behavior have been turned upside down.

Important aspects of social behavior seem, at times, to have improved—less nonessential travel, less consumption, more family focus, reduced greenhouse gas pollution (17 percent less worldwide in early April 2020), increase in demand for healthy, often organic, home-cooked foods, more interest in natural health remedies, appreciation for nature, mutual aid, and more attention paid to the plight of nursing home patients, farmworkers, small farmers, health care workers, and food chain workers.

Unfortunately, most of the impacts of the pandemic have been quite negative, indeed catastrophic: a massive number of hospitalizations and deaths attributed to COVID-19, widespread anxiety and fear, extreme political polarization, media censorship, draconian lockdowns, closures of schools and businesses, and economic meltdown, including increasing numbers of bankruptcies of small, medium-sized, and even large businesses, with 30 million workers unemployed in the US alone.

As of January 20, 2021, the US Centers for Disease Control (CDC) reports that *just over 400,000 Americans have died either from or with COVID-19*, a yearly average of 1,096 per day.[5] However, CDC data released August 26, 2020, show that only 6 percent of COVID-19 deaths in the US had COVID-19 listed as the sole cause of death on the death certificate. The remainder, 94 percent, had an average of 2.6 comorbidities or additional causes of death.[6]

The overwhelming majority of COVID-19's victims in the US (80 percent) have been elderly (65 years or older),[7] with almost all suffering from serious preexisting chronic diseases or medical conditions, and almost half of all deaths taking place in nursing homes.[8] Global deaths in 2020 from or with COVID-19 are estimated to be 2.8 million,[9] with economic damage estimated at $16 trillion.[10]

Millions of Americans (and hundreds of millions worldwide)—especially lower-income workers—have lost their jobs and livelihoods, and scores of thousands of US businesses, especially small businesses, have gone under. While 40 percent of Americans are unable to afford an emergency $400 bill,[11] many large transnational corporations (Amazon, Big Pharma, Walmart,

McDonald's, and more) and billionaires such as Bill Gates, Jeff Bezos, and Mark Zuckerberg have prospered. According to a study by the Institute for Policy Studies, the combined wealth of US billionaires surpassed $1 trillion in gains during the height of the pandemic.[12]

An entire generation of children and students have had their education and lives turned upside down. For most of the world's seven billion people, COVID-19 has been the most destructive event in their lifetime, a turning point in world history.

The Rise of a Digital Dictatorship

Driven by fear, confusion, and manufactured consent, the US, and indeed much of the world, seems to be degenerating into what Indian scholar, author, and environmental activist Dr. Vandana Shiva and others have described as a digital dictatorship.

This 21st-century digital dictatorship is thus far most advanced in China, the world's largest and fastest-growing economy, with its militarized and authoritarian regime of surveillance, centralized planning, censorship, and total control. But a Westernized, globalist model of elite manipulation and control is also emerging that competes, but also cooperates, with the Chinese elite.

This Western elite is led by hypercapitalist billionaires such as Bill Gates (Microsoft) and Eric Schmidt (Google), along with their other tech giant cohorts in Silicon Valley (Facebook, Amazon, Apple, Oracle, et cetera), Big Pharma, Wall Street, multinational corporate executives, the World Economic Forum, and the military-industrial biowar complex.

This global elite, aided and abetted by indentured politicians, scientists, media moguls, and government bureaucrats, is maneuvering to use the current pandemic and economic meltdown to grab unprecedented power and wealth (the "Shock Doctrine" as Naomi Klein has termed it) and impose, in the name of public health and "biodefense," draconian surveillance, censorship, and control.

The global elite's unprecedented power grab includes eliminating the last vestiges of participatory democracy, free speech, cultural diversity, ecological biodiversity, and individual freedom.

In a May 2020 article for *The Intercept*, Naomi Klein offers up a preview of this emerging dystopia, which she calls a "Screen New Deal":

> *. . . a future in which our homes are never again exclusively personal spaces but are also, via high-speed digital connectivity, our schools, our doctor's offices, our gyms, and, if determined by the state, our jails . . . in*

the future under hasty construction, all of these trends are poised for a warp-speed acceleration.

This is a future in which, for the privileged, almost everything is home delivered, either virtually via streaming and cloud technology, or physically via driverless vehicle or drone, then screen "shared" on a mediated platform. It's a future that employs far fewer teachers, doctors, and drivers. It accepts no cash or credit cards (under guise of virus control) and has skeletal mass transit and far less live art. It's a future that claims to be run on "artificial intelligence" but is actually held together by tens of millions of anonymous workers tucked away in warehouses, data centers, content moderation mills, electronic sweatshops, lithium mines, industrial farms, meat-processing plants, and prisons, where they are left unprotected from disease and hyperexploitation.

It's a future in which our every move, our every word, our every relationship is trackable, traceable and data-mineable by unprecedented collaborations between government and tech giants.[13]

And how do these digital overlords and billionaires expect to convince us to give up our basic freedoms and democratic rights and become loyal serfs of a Great Reset and a New World Order? By taking advantage of the fear, helplessness, division, and confusion encompassing the world—spreading disinformation, promoting panic, offering false cures through Big Pharma vaccines and drugs, and ultimately by dividing and conquering the general public.

It is time to ask ourselves some very pressing questions. Do pandemic response measures such as lockdowns, mask mandates, social distancing, and quarantine regulations serve to protect the world's population from COVID-19, or do these measures serve only to increase fear and thereby facilitate compliance with tyrannical liberty-eroding edicts?

More people are now starting to wake up to the fact that the restrictions put into place under the guise of protecting public health are anything but temporary and will likely remain permanent. They're part of a much larger long-term plan, and the end goal is to usher in a new way of life with radically reduced previous freedoms. This means that, eventually, everyone must decide which is more important: personal liberty or false security?

How can we preserve consumer choice and health freedom and promote regenerative food, farming, natural health, and participatory democracy? How can we overcome fear, protect ourselves, our families, and our loved ones from suffering from chronic disease, COVID-19, and even the threat of future pandemics?

Why We Are Writing This Book

The reason we are writing this book is because we believe the COVID-19 pandemic can become, not a dystopian dead end, but a portal to a better world. The current crisis is alarming, but it offers us an opportunity to qualitatively improve public health and planetary health and regenerate the global grassroots. We can cross over from what can only be described as the Disease of Nations to true health and democracy.

We believe it is possible to educate and empower the average person, and defeat the digital dictators, fearmongers, mad scientists, medical fascists, and indentured, bought-and-paid-for politicians. Individually and, most important, working together, we can head off the looming threat of the digital dictatorship and so-called Great Reset that is already emerging. We can take control of our health, our communities, and our destiny.

We believe that the biotechnocrats, the military, and the transnational economic elite hell-bent on global domination have overreached themselves. In the midst of an unprecedented global disaster and government failure to solve the COVID-19 crisis, it's time for us to take matters into our own hands.

Now is the time for a global awakening. Now is the time for a local-to-global resistance.

Reckless Science and Bioweapons

With mounting evidence and increasing certainty, a growing number of independent scientists, investigators, and lawyers have begun to deconstruct and critique the "official story" on the origins, nature, dangers, prevention, and treatment of the COVID-19 pandemic.[14]

The "official story," dogmatically upheld by the Chinese government and military, Big Pharma, Bill Gates, the US Centers for Disease Control, the National Institutes of Health (NIH), the mass media, and the tech giants, is that the SARS-CoV-2 virus emerged "naturally" from nature and then inexplicably jumped the species barrier from bats into humans, precipitating the most serious and deadly epidemic since the Spanish flu one hundred years ago, which infected one-third of the world's population at the time and killed up to 50 million people.

According to establishment virologists and gene engineers (who get their money from military biodefense programs, government funding, and Big Pharma), a relatively innocuous and heretofore non-contagious coronavirus quickly mutated into a deadly killer, leaving behind no biological or epidemiological traces whatsoever of its rapid evolution.[15]

Moreover, in a billion-to-one coincidence, this deadly viral mutation and ensuing epidemic emerged in the exact same densely populated urban neighborhood (hundreds of miles from the nearest bat cave) in Wuhan, China, where a series of controversial genetic engineering experiments involving the weaponization (euphemistically called gain-of-function experimentation[16]) of coronaviruses were being conducted in several badly managed, accident-prone labs.[17]

The powers that be in Beijing and Washington like to reassure us that researchers in places like the Wuhan Institute of Virology, the Wuhan Center for Disease Control and Prevention, the US Army Biological Warfare Laboratories at Fort Detrick, Maryland, the University of North Carolina, and the Johns Hopkins Center for Health Security, are only "studying" (not manipulating or weaponizing) dangerous pathogens like bat coronaviruses, and that security in these government/WHO/NIH-monitored labs is so strict that accidents could never happen.

But a number of well-respected scientific critics of genetic engineering and biological warfare have been sounding the alarm for decades.

Critics including Francis Boyle (author of the 1989 US bioterrorism law banning bioweapons research) and Dr. Richard Ebright of Rutgers University's Waksman Institute of Microbiology, along with hundreds of other scientists, have warned that experiments and manipulations of viruses and other pathogens are inherently extremely dangerous (not to mention that they violate international law), given human error and the fact that security has been dangerously lax in the world's biowarfare/biodefense laboratories.[18]

Among the proponents of the official story that SARS-CoV-2 arose "naturally" are the Big Pharma–affiliated EcoHealth Alliance,[19] as well as a secretive and little-known network of US military biowarfare/biodefense funders including Defense Advanced Research Projects Agency (DARPA) and the Office of the Assistant Secretary of Preparedness and Response (ASPR) for the US Department of Health and Human Services.[20]

America's military/pharma complex has provided major funding for the Wuhan lab, the University of North Carolina (where scientists have weaponized SARS viruses), the Fort Detrick, Maryland, military chemical and biological weapons lab, as well as several hundred other biomedical/biowarfare labs across the world.[21]

Another vocal proponent of the official story is the World Health Organization (WHO), the agency that was supposedly monitoring the accident-prone Wuhan lab.[22] WHO's major funders include China, the US government, and Bill Gates, along with Big Pharma drug and vaccine manufacturers.

Biowarfare: Weaponizing Viruses

Despite an ongoing cover-up by Chinese and US government authorities, the biotech industry, Big Pharma, the military-industrial complex, and the mass media, there is growing scientific consensus that the COVID-19 virus was created and (*most likely accidentally*) leaked from a dual-use military/civilian lab in Wuhan, China.[23]

Unbeknownst to the public, a shadowy international network of thousands of virologists, gene engineers, military scientists, and biotech entrepreneurs are weaponizing viruses, bacteria, and microorganisms in civilian and military labs under the euphemistically called gain-of-function research.

They hide behind the guise of biodefense, biomedicine, and vaccine research. But as investigative reporter and bioweapons expert Sam Husseini writes, gain-of-function/biowarfare scientists in labs such as Wuhan and Fort Detrick are deliberately and recklessly evading international law:

> *Governments that participate in such biological weapon research generally distinguish between "biowarfare" and "biodefense," as if to paint such "defense" programs as necessary. But this is rhetorical sleight-of-hand; the two concepts are largely indistinguishable.*
>
> *"Biodefense" implies tacit biowarfare, breeding more dangerous pathogens for the alleged purpose of finding a way to fight them. While this work appears to have succeeded in creating deadly and infectious agents, including deadlier flu strains, such "defense" research is impotent in its ability to defend us from this pandemic.*[24]

A growing arsenal of Franken-viruses and microorganisms have been created, despite US and international laws supposedly banning biowarfare weapons and experimentation.[25] Over the past three decades, a disturbing number of these so-called dual-use biowarfare/biodefense labs have experienced leaks, accidents, thefts, and even deliberate releases like the 2001 anthrax attacks in the US.[26]

Because the SARS-CoV-2 virus is so infectious and dangerous, the pandemic fearmongers tell us, there are currently no drugs, treatment protocols, supplements, natural herbs, dietary, or natural health practices that can strengthen our natural immune systems and protect us from serious illness, hospitalization, or even death from the virus.

We have no alternative, whether we are young or old, healthy or seriously health-impaired, but to wear masks, not just in enclosed public spaces but everywhere. In addition, we must wash our hands incessantly, stay six feet or more apart, and shut down schools, social gatherings, churches, businesses, and entire economies.

We have no choice, Big Government and Big Pharma tell us, but to stay home, obey authority, and wait for Big Pharma or the Chinese government to deliver a "cure," a magic vaccine that has been inadequately tested, rushed to market, likely genetically engineered, and designed to maximize corporate profits.

Medical Malpractice

Let's not forget that after decades of massively funded research, Big Pharma has never been able to develop an effective vaccine for a coronavirus. A genetically engineered vaccine designed to modify (perhaps permanently) human RNA has never been allowed on the market, in part because a number of these coronavirus vaccines seem to create dangerous ADE (antibody-dependent-enhancement) side effects in many of those injected, especially the elderly, making them more susceptible to dangerous disease.[27]

And what about the safety record[28] and reckless, liability-free practices of the Big Pharma corporations that produce vaccines (Merck, AstraZeneca, Johnson & Johnson, BioNTech, GlaxoSmithKline, Pfizer, and so on)?[29]

Then we all have to contend with the mass media, Big Pharma, the WHO, and the tech giants that are censoring information about successful treatments carried out by doctors across the world using low-cost but effective drugs and supplements like quercetin and zinc (the Swiss Protocol),[30] hydroxychloroquine (at proper low doses with zinc supplements and the antibiotic azithromycin),[31] ivermectin[32] (which appears particularly effective at preventing SARS-CoV-2 infection), vitamin D supplementation,[33] nebulizing hydrogen peroxide into your sinuses, throat, and lungs,[34] and the "COVID-19 Critical Care"[35] also known as the MATH+ protocol, both for prevention and for treatment for those hospitalized.[36]

It's important to be aware that a growing number of these Big Pharma giants are already selling billions of dollars of COVID-19 vaccines to governments and the military in secretive, no-bid contracts,[37] even though none of these vaccines are being properly safety-tested before being rubber-stamped as safe and effective.

Would-be digital dictators like Bill Gates,[38] Silicon Valley surveillance capitalists,[39] and indentured, Pharma-funded politicians are floating proposals for mandatory vaccinations, injectable medical biosurveillance computer chips, mandatory tracing, vaccine passports, and elimination of basic constitutional rights.[40]

Biowarfare gene engineers and lab technicians, hiding behind the excuse of biomedicine and vaccine research, are, at this very moment, weaponizing new viruses and bacteria (including combining deadly anthrax bacteria with SARS-CoV-2 and aerosolizing the bird flu) in basically nonregulated and accident-prone labs.[41]

And finally, there are massive financial conflicts of interest and increasing violations of free speech that the major media networks and the internet giants Facebook, Google, Amazon, and their subsidiaries are carrying out, marginalizing, or totally censoring alternative information about the origins, nature, prevention, and treatment of COVID-19.[42]

Junk Food, Environmental Pollution, and Chronic Disease

The shocking truth is starting to come out about the real origins of COVID-19.[43] But perhaps more shocking still is the way that this disease has shone a light on the fragility of our food system, the lack of transparency in our regulatory and scientific communities, and the terrifying vulnerabilities of the human body, worn down by a lifetime of junk food and exposure to toxic chemicals.

The public health bottom line is that the SARS-CoV-2 virus is not so much a deadly plague in itself, but rather a viral trigger that aggravates and magnifies preexisting chronic medical conditions, what pathologists call comorbidities. And, of course, the majority of these comorbidities are diet- and food-related. Others are caused by our exposure to toxic chemicals, electromagnetic radiation,[44] and other environmental contaminants.

According to the CDC, 94 percent of death certificates for COVID-19 victims in the US list a number of comorbidities or underlying health co-factors in their deaths, including diabetes, obesity, heart disease, lung disease, kidney disease, dementia, and hypertension.[45]

According to the *New York Times*:[46]

> *The correlations between Covid-19 and obesity are worrisome. In one report published last month, researchers found that people with obesity who caught the coronavirus were more than twice as likely to end up in the hospital and nearly 50 percent more likely to die of Covid-19.[47] Another study, which has not yet been peer-reviewed, showed that among nearly 17,000 hospitalized Covid-19 patients in the United States, more than 77 percent had excess weight or obesity.[48]*

Unfortunately, many people, especially the elderly in nursing homes and in hospitals, are not in good health. The virus sickens and kills elderly people in poor health, as well as at-risk adults, especially those in low-income communities suffering from chronic disease, polluted air and water, poor diets, and limited access to healthy foods, nutritional supplements, and natural health information and treatments.

SARS-CoV-2 targets those, especially older adults, who have been damaged by long-term exposure to industrial food and pollution. Victims of standard grocery store and restaurant fare, stuffed with Big Food/Big Ag carbs and calories, are left undernourished and metabolically unbalanced by the typical American diet.

These victims-in-waiting are typically suffering from a variety of chronic diseases (especially obesity, diabetes, and high blood pressure), weakened immune systems, low vitamin D levels, and poor gut and digestive health.

The primary reason why so many Americans are chronically ill is that Big Food and Big Ag in the US (and across the world) basically produce—and in fact are subsidized by governments to produce—what can only be described as junk food commodities. These junk foods and beverages, which make up 60 percent or more of the calories in the typical American diet, are highly processed, sugar- and carb-laden, and laced with pesticide, antibiotic, and chemical residues. In toxic combination with the typical American's overconsumption of factory-farmed meat and animal products, US junk food diets are a virtual prescription for chronic disease and premature death.[49]

One major reason for the prevalence of junk foods in the modern diet is that these foods, at least at the grocery checkout counter or fast-food register, are cheap.

Typically, junk food sells at one-quarter of the price per calorie of real whole foods (vegetables, fruits, grains), with the true costs of production and consumption, including damage to public health, the environment, and the climate, concealed from the public.

Junk foods and sodas are manufactured to be tasty and addictive, cheap and plentiful, but they are ultimately poisonous. They certainly can quickly and conveniently fill your belly, especially if you're operating on a limited budget, but they can make you fat, clog your arteries, and lead to cancer, heart disease, and dementia. Junk foods destroy your health and damage your gut biome and immune system, setting you up for chronic disease and viral triggers such as COVID-19 that can aggravate and magnify existing disease.

In a society that would dare to put public health ahead of corporate profits, junk food would be banned or so heavily taxed (like tobacco) that it would be displaced by real food. Big Food and Big Ag would collapse and Big Pharma's drug profits would shrink drastically. Organic and regenerative food and farming would become the norm for everyone, instead of the alternative.

While acknowledging that we have to stop the reckless military/scientific genetic engineering that brought on this pandemic and global economic meltdown, media censorship, and suspension of fundamental democratic rights, we also need to defend ourselves and our families by changing our diets and collectively moving away from the industrialized, degenerate food and farming system that sets up people for premature death and hospitalization.[50]

The reason so many people are dying from or with COVID-19 and other preventable chronic diseases listed on their death certificates is that the US has one of the unhealthiest populations in the industrialized world.[51]

The "cure" for chronic disease and premature death, the most important preventive step to fight off invading viruses, is organic and regenerative food, complemented by appropriate nutritional supplements, herbs, and natural health remedies.

Chronic disease and comorbidities can not only be prevented and mitigated, but also cured, especially if we as a society prioritize healthy food, exercise, and nutritional supplementation, and clean up the environment. But right now our nursing homes, health care clinics, hospital facilities, and institutional settings are doing just the opposite.

We need to avoid subsidizing junk food and direct those funds to healthy, organic food for everyone, young and old, rich and poor. And we need a fundamental change in medical priorities, from just treating chronic disease with pharmaceutical drugs to preventing chronic disease with healthy "food as medicine" as well as other natural health promoters such as medicinal herbs, vitamins, and nutritional supplements. This is how we can defeat COVID-19 and the epidemic of obesity, diabetes, cancer, heart disease, and other chronic diseases.

We also need to identify and bring to justice the criminal perpetrators of COVID-19, and ban biological warfare experiments, such as those weaponizing viruses, forever.

But at the same time, we must educate the public that "preexisting" business-as-usual practices and policies (bad food, air and environmental pollution, pesticides, and contaminated vaccines) are the real deadly drivers of this pandemic and, along with it, the lockdown and meltdown of the global economy.

We need to educate others that even as Facebook and the mass media censor truth,[52] eating healthy food, strengthening our immune systems, getting plenty of fresh air and sunshine, and exercise are our best defenses against COVID-19 and the epidemic of chronic diseases that have undermined public health.

The body politic, though still divided among those who live in fear of COVID-19, those who worry about how they are going to survive economically, and those who have reached a psychological breaking point after being quarantined and socially isolated, can still be united.

We can move forward together and put this crisis behind us *if* we can freely share information and experiences and get to the truth about how this pandemic started, who is lying to us, who is trying to manipulate and control us, and how, building upon the positive preventive and therapeutic solutions that have actually worked in certain areas across the world, we can move beyond this nightmare.

We must stop fighting among ourselves—Democrats, independents, and Republicans; liberals and libertarians; radicals and conservatives—and instead focus on the fundamental ethical values and social goals that unite us. We must strive to imagine, and then build, from the ruins of the old, a new world.

Together we can move beyond fear and doom and gloom. As renowned Indian activist Vandana Shiva explained in a recent interview, "We have to resist fear and we have to resist hate . . . we have to absolutely not become victims of fearmongering . . . we don't have the luxury to be hopeless . . . To be alive today means hope is something you must cultivate on a daily basis. Cultivating hope is cultivating resistance."[53]

Together as a local-to-global community, we can share and implement the positive solutions to our deteriorating public health and the disease of nations. These positive solutions already exist—healthy, organic, and regenerative food, farming, and land use; renewable energy and a clean environment; natural and integrative health practices; peace, justice, and participatory democracy.

But to step through the portal of pandemic and fear, we must stop obsessing and arguing over our secondary differences, and focus instead on what we all support: healing and regenerating the body politic and planetary health.

As fellow human beings on a planet in crisis, we must avoid the trap of magnifying our differences, of treating one another as enemies. As Robert Kennedy, Jr., reminds us: "The enemy is Big Tech, Big Data, Big Oil, Big Pharma, the medical cartel, the government totalitarian elements that are trying to oppress us, that are trying to rob us of our liberties, of our democracy, of our freedom of thought, of our freedom of expression, of our freedom of assembly and all of the freedoms that give dignity to humanity."[54]

Let's now look together at what really happened and what's continuing to really happen with COVID-19: its origins, nature, virulence, threats, prevention, and treatment. And let's try to strategize how we can defeat the current attempt at a coup d'état by the global elite and the digital dictators and build a new future that is healthy, regenerative, just, participatory, and democratic.

CHAPTER TWO

Lab Leak or Natural Origin?

By Ronnie Cummins

David A. Relman, a Stanford University microbiologist, writes in the Proceedings of the National Academy of Sciences, *"the 'origin story' is missing many key details," including a recent detailed evolutionary history of the virus, identity of its most recent ancestors and surprisingly, the place, time, and mechanism of transmission of the first human infection.*

—Editorial Board of the *Washington Post*,
November 14, 2020[1]

For almost 30 years a growing number of scientists and activists, including the authors of this book, have tried to warn the world about the inherent dangers of "playing God," meaning genetically engineering DNA, the building blocks of life, and now messenger RNA (in the new experimental COVID vaccines, for example).

A big part of the reason why playing God is so dangerous is that these aims are pursued with little or no government regulation or consideration for the potential dangers of genetically modified organisms (GMOs) to human health and the environment.

As a result of our educational efforts, many, if not most, consumers around the world have become wary of genetically engineered foods and crops, as well as the toxic chemicals such as Bayer/Monsanto's glyphosate/Roundup pesticide that always accompany these GMOs. Unfortunately, most people have not heard much about the other branch of genetic engineering and gene editing—the secretive and shadowy world of bioweapons, biosafety, and biomedical research.

In this high-tech world of biotechnocracy, thousands of global scientists and researchers, funded by Big Pharma and the military-industrial complex, are genetically engineering viruses, bacteria, and microorganisms to make them more infective, virulent, and dangerous.

The biotechnocracy hides behind the excuse that these are not bioweapons, which are supposed to be banned by an international Biological and Toxin Weapons Convention, but rather are biomedical or biosafety experiments designed to help humankind develop new drugs or vaccines to fight off epidemics and disease.[2]

Unfortunately, after 30 years of this so-called biosafety research, there have been hundreds of documented leaks, thefts, accidents, and even deliberate releases (such as the 2001 anthrax attacks in the US) every year. This doesn't even include unreported incidents involving weaponized viruses and bacteria. But absolutely no new effective vaccines or drugs have come out of this dangerous gain-of-function research.[3]

While gain of function may sound like a positive thing, it actually refers to the act of weaponizing viruses, often through genetic manipulation. Coronaviruses like SARS typically have a narrow host range, infecting one or just a few species, such as bats. However, using targeted RNA recombination, gene engineers can manipulate viruses such as COVID-19 for "gain of function" to enable them to infect other species (human cells), interfere with immune system response, and readily spread through the air.[4]

The Official Story Is Crumbling

The official story, which claims that the SARS-CoV-2 virus is natural as opposed to lab-engineered, is slowly but surely crumbling. A growing number of independent scientists and investigators are exposing the factual errors and outright lies of the establishment narrative by analyzing a growing body of evidence and publishing their findings, in spite of widespread censorship by scientific journals, the mass media, and the internet giants.

Among the international critics publicly deconstructing the official story are hundreds of respected scientists and researchers, including: Chris Martenson, Alina Chan, Meryl Nass, Moreno Colaiacovo, Richard Ebright, Nikolai Petrovsky, Etienne Decroly, David Relman, Milton Leitenberg, Stuart Newman, Aksel Fridstrøm, Nils August Andresen, Rossana Segreto, Yuri Deigin, Jonathan Latham, Alison Wilson, Vandana Shiva, Sam Husseini, Luc Montagnier, Carey Gillam, Claire Robinson, Jonathan Matthews, Michael Antoniou, Joseph Tritto, Lynn Klotz, Filippa Lentzos, Richard Pilch, Miles Pomper, Jill Luster, Birger Sørensen, Angus Dalgleish, Andres Susrud, Monali Rahalkar, Rahul Bahulikar, and many others.[5]

Upholding the official story and opposing these critics, often denouncing them as "conspiracy theorists," the Chinese government, the US government,

Big Pharma, Silicon Valley, and the global biomedical/biodefense elite argue that the SARS-CoV-2 coronavirus, a distant relative of the virulent but far less infectious SARS-CoV virus that infected 8,000 people in 2002–2004 and then disappeared, arose naturally from a wild bat.

This wild bat virus then somehow evolved, recombining its genes with another wild animal (a type of anteater called a pangolin), making it more infectious and virulent.[6] Following this miraculous recombination event, SARS-CoV-2 then mutated again, gaining the ability to infect humans, causing a global pandemic, while leaving no biological, genomic, or epidemiological traces of its evolutionary history in its wake.

If there is indeed a natural explanation for the origins of COVID-19, all that the Chinese (and the US) government and the research scientists involved have to do is to come forward and provide us with the evidence.

This evidence, as Dr. David Relman points out in the *Proceedings of the National Academy of Sciences*, would need to provide laboratory samples and scientific data showing the identity of SARS-CoV-2's most recent ancestors, as well as "the place, time, and mechanism of transmission of the first human infection."[7]

But don't hold your breath waiting for this to happen, since no one wants to take the blame, and pay the liability, for the most destructive lab accident or lab release in human history.

The "natural origins" story has been regurgitated so many times by the mass media, the internet giants, the science and medical establishment, and government authorities that it's no wonder the majority of ordinary people remain confused, misinformed, and fearful.

Anyone daring to challenge the official narrative, pointing to the preponderance of evidence that SARS-CoV-2 appears to be a synthetic construct that was genetically engineered in a biomedical/bioweapons laboratory in Wuhan, China, and released, either accidentally (most probably) or deliberately, is berated and dismissed as a "conspiracy theorist" and then typically summarily censored and/or banned on their social media platforms.

As the official narrative goes, SARS-CoV-2 arose suddenly, without warning, like a plague, almost like an act of God, allegedly originating and spreading from the Huanan Seafood Wholesale Market, a "wet market" in Wuhan where exotic live animals such as bats and pangolins were supposedly sold, killed, and eaten.

Needless to say, the lurid image of a Chinese market vendor and their customers at the Wuhan Seafood Market killing a bat, eating it, and then contracting a dreadful disease was scary and repulsive, like something out of a horror movie. It also smacks of racism, as it plays on stereotypes of Asian people eating foods that are foreign—and therefore repellent—to Western palates.

Most of the world media, as well as a group of elite scientists who experiment on bats and viruses, repeated the Chinese government's "bat-in-the-wet-market story" ad nauseam. Unsurprisingly, a global cry rose up to ban wet markets and bush meat.

The Great Cover-Up

In order to conceal their scientific malpractice and criminal negligence and protect their "right" to carry out dangerous, unregulated research in violation of a global treaty banning bioweapons research, and to safeguard billions of dollars in annual Big Pharma and GMO industry profits, Chinese and US officials, Big Pharma, Facebook, Google, and an arrogant and unscrupulous network of global scientists have been frantically trying to cover up the lab origins of the COVID-19 pandemic.

From the beginning, Chinese government officials have lied and tried to conceal the facts surrounding COVID-19, aided and abetted in many cases by the World Health Organization (of which China and Bill Gates are major funders—something we'll cover more in chapter 3), and an indentured network of gene engineers and virologists from the US and other countries who study and weaponize viruses and bacteria under the guise of biomedicine or vaccine research.

The first stage of the cover-up involved trying to conceal or delay admitting that a new SARS-like epidemic had emerged in Wuhan, and that, unlike the first SARS outbreak of 2002–04 in China, this one was highly transmissible. Although the world media dutifully repeated the later-discredited explanation that COVID-19 came from a wet market in Wuhan, a number of news organizations did begin to investigate and expose some of the early duplicity, stalling, and lies of the Chinese government.

These outlets began reporting the fact that the Chinese government delayed for a month or more admitting that a new, previously unknown, and serious respiratory outbreak, COVID-19, had emerged in November–December 2019. Chinese officials (and the WHO) then delayed letting the world know that this SARS-CoV-2 virus was spreading rapidly person-to-person in Wuhan until February 19, 2020, despite warnings from some of their top scientists.

The Chinese government meanwhile was censoring and repressing scientists and doctors who were trying to get out the word that a serious health crisis was emerging. As Canadian journalist Andrew Nikiforuk pointed out: "Faced with the coronavirus threat, Chinese authorities, according to comprehensive reports by the *Wall Street Journal* and the *New York Times*, suppressed whistleblowers, ignored critical evidence and responded so tardily to the outbreak that they moved to compensate for their failures with a draconian lockdown."[8]

Little noticed by the world press, however, were the growing signs of another, even more insidious cover-up—that SARS-CoV-2 didn't emerge naturally, but in fact had escaped or been released from one of Wuhan's two biosafety research labs, where gain-of-function experiments were taking place in badly managed, accident-prone facilities. In these labs, unbeknown to the world, thousands of bat coronaviruses were stored and in some cases had been weaponized.

Lab Leaks of Dangerous Viruses and Bacteria

A growing arsenal of synthetic viruses have been engineered in so-called dual-use biowarfare/biodefense labs, despite US and international laws banning biowarfare weapons and experimentation.[9] A disturbing number of these labs have experienced leaks, accidents, and thefts over the past three decades.

As the well-respected *Bulletin of the Atomic Scientists* recently warned: "A safety breach at a Chinese Center for Disease Control and Prevention lab is believed to have caused four suspected SARS cases, including one death, in Beijing in 2004. A similar accident caused 65 lab workers of Lanzhou Veterinary Research Institute to be infected with brucellosis in December 2019 . . . In January 2020, a renowned Chinese scientist, Li Ning, was sentenced to 12 years in prison for selling experimental animals to local markets."[10]

China is hardly the only place to tolerate this type of reckless scientific experimentation and then experience such accidents. Time and again serious safety breaches have been identified at laboratories working with the most lethal and dangerous pathogens in the world.[11] A 2016 *USA Today* investigation, for instance, revealed an incident involving cascading equipment failures in a decontamination chamber as US Centers for Disease Control and Prevention (CDC) researchers tried to leave a biosafety level 4 lab. The lab likely stored samples of the viruses causing Ebola and smallpox, according to the report.[12]

In 2014 six glass vials of smallpox virus were accidentally found in a storeroom in the US Food and Drug Administration's lab at the National Institutes of Health.[13] It was the second time in one month that mishandling of potential deadly infectious agents was exposed. Shortly before this shocking discovery, the CDC realized that staff had accidentally sent live anthrax between laboratories, exposing at least 84 workers. In an investigation officials found other mishaps that had occurred in the preceding decade.[14]

The next year the Pentagon realized a Dugway Proving Ground laboratory had been sending incompletely inactivated anthrax to 200 laboratories around the world for the past 12 years. According to a Government Accountability Office (GAO) report issued in August 2016, incompletely inactivated anthrax was sent out on at least 21 occasions between 2003 and 2015.[15]

In 2017 the BSL 4 lab on Galveston Island was hit by a massive storm and severe flooding, raising questions about what might happen were some of the pathogens kept there to get out.[16] In August 2019 the US Army Fort Detrick, Maryland, Biological Warfare Lab was temporarily shut down for improper disposal of dangerous pathogens, according to a *New York Times* report. Officials refused to provide details about the pathogens or the leak, citing "national security" concerns.[17]

In 2017 Tim Trevan, a Maryland biosafety consultant, expressed concern about viral threats potentially escaping the Wuhan National Biosafety Laboratory specifically.[18] US diplomatic cables sent in 2018 also warned about "possible safety breaches at a lab in Wuhan."[19]

Certainly the US military and the CIA, and their Chinese counterparts, along with other nations funding bioweapons experiments under the guise of biosafety or vaccine research, do not want to admit that the official story on the natural origins of COVID-19 is based upon propaganda rather than fact.

The global biological arms race has now literally blown up in our faces. Nor did Donald Trump or Dr. Anthony Fauci and the NIH want the true story to come out, since the NIH-funded EcoHealth Alliance was providing money for the reckless gain-of-function research experimentation on coronaviruses at the Wuhan lab.

According to *Newsweek*, this research was conducted in two parts.[20] The first, which began in 2014 and ended in 2019, focused on "understanding the risk of bat coronavirus emergence."[21] Initial findings were published in *Nature Medicine* in 2015.[22] The program, which had a budget of $3.7 million, was led by Wuhan virologist Shi Zhengli and sought to catalog wild bat coronaviruses. It also involved US scientists from the University of North Carolina and Harvard.[23]

The second phase, which began in 2019, included additional surveillance of coronaviruses along with gain-of-function research to investigate how bat coronaviruses might mutate to affect humans. This second phase was run by EcoHealth Alliance under the direction of its president, Peter Daszak, an expert on disease ecology.

But lest we blame it all on the Chinese military or the Trump administration, we should keep in mind that this incredibly dangerous mad science,
being funded and carried out right now at places like Fort Detrick, Columbia
University, and the University of North Carolina by the US government and
the military, has been going on ever since World War II. It's been funded
and carried out in turn by the Truman, Eisenhower, Kennedy, Nixon, Carter,
Reagan, Bush Sr., Clinton, Bush Jr., and Obama administrations.[24]

Second Stage of the Cover-Up

The second stage of the cover-up, unfortunately ignored by most of the media,
has included the systematic destruction of forensic evidence, including:

- Tests and samples taken from the Wuhan market and nearby labs in
  late December 2019.[25]
- The takeover of the Wuhan Institute of Virology by the Chinese
  military and its top biological weapons specialist on January 26, 2020.[26]
- The removal from public online data bases of 20,000 bat virus genomes
  collected by the Wuhan Institute of Virology and others.[27]
- The censoring and even disappearing of Chinese scientists who pointed
  out that SARS-CoV-2 was likely a lab release with the potential to set
  off a dangerous pandemic.
- A ban on publishing any articles on SARS-CoV-2 without the formal
  prior approval of the Chinese military.
- The secret alteration of data sets in published papers without publishing notices of correction.[28]
- The deletion of an estimated 300 coronavirus studies from the state-run
  National Natural Science Foundation of China's database in January
  2021, including studies conducted at the Wuhan Institute of Virology.[29]

In addition to the destruction of evidence, a manufactured narrative that
dismissed the lab-origin theory offhand was put into place early on, and has
been adhered to ever since. Peter Daszak of the EcoHealth Alliance, which
funneled NIH grant money over to the Wuhan Institute of Virology for coronavirus research, played a key role in that scheme.

On February 18, 2020, *The Lancet* published a scientific statement signed
by 27 researchers denouncing theories that COVID-19 came from a lab,
stating, "We stand together to strongly condemn conspiracy theories suggesting that COVID-19 does not have a natural origin. Scientists from multiple
countries have published and analyzed genomes of the causative agent . . .

SARS-CoV-2, and they overwhelmingly conclude that this coronavirus originated in wildlife . . ."[30]

Social media fact checkers routinely relied on this statement to censor evidence showing SARS-CoV-2 appears human-made. As it turns out, Daszak was key orchestrator of this plot to dissuade public discussion about the virus's origin.[31] He drafted the statement, but emails obtained through the Freedom of Information Act (FOIA) reveal he did not want it to be "identifiable as coming from any one organization or person."[32] Rather, he wanted it to appear as "a letter from leading scientists."[33]

As 2021 rolled around and calls for a real investigation into the origin of SARS-CoV-2 heated up, Daszak was then assigned to not just one but two separate commissions tasked with this endeavor: *The Lancet*'s COVID-19 Commission[34] and the World Health Organization's investigative committee.[35]

What are the chances that Daszak—who played a key role in manufacturing the narrative that SARS-CoV-2 is zoonotic in origin to begin with—will come to any other conclusion at the end of these investigations? Five other members of the *Lancet* Commission also signed the February 18, 2020, statement in *The Lancet*, which puts their credibility in question as well.

Interestingly, while accusations of a lab escape continue to be denied by Chinese authorities, biomedical/biodefense labs in China, including the Wuhan Institute of Virology and the Center for Disease Control, have implemented new laboratory procedures to tighten up security and safety measures in the wake of the pandemic.[36]

The Collapse of the Bat-in-the-Market Origins Story

For months after the onset of the epidemic, the mass media basically ignored a number of scientific and media reports that no animals at the Wuhan market had tested positive for COVID-19, and that approximately one-third of the initial group of reported human COVID-19 cases in Wuhan from early December 2019 had no connection to the seafood market whatsoever, including the first reported case.[37]

The media also largely ignored the testimony of numerous people from the Huanan Seafood Market who categorically stated that there were no bats for sale, and none eaten, in the market. In fact, the nearest bat caves to the Wuhan Market were 600 miles away, and when SARS-CoV-2 emerged in Wuhan, these wild bats were in hibernation.

In January 2020 a Beijing newspaper had reported that "patient zero," the first victim of the COVID-19 virus, was in fact Huang Yanling, a scientist at the Wuhan Institute of Virology, and although this report has since been removed from the internet, the rumors persist.[38] In February well-respected Chinese scientists Botao Xiao and Lei Xiao pointed out in a pre-publication scientific article (quickly suppressed and removed from the internet by the Chinese government): "According to municipal reports and the testimonies of 31 residents and 28 visitors, the bat was never a food source in the city, and no bat was traded in the market."[39]

Although there were apparently no wild bats for sale or consumed in the Wuhan market, there were many bats and bat viruses being stored and experimented on in two supposedly high-security research labs nearby.

One of these, the Chinese Center for Disease Control, is just 300 yards away from the Huanan Seafood Market and adjacent to Union Hospital, where several early cases of COVID-19-infected hospital doctors were reported. Another, the Wuhan Institute of Virology, is seven miles away. The conclusion of Botao Xiao and Lei Xiao, suppressed by the Chinese government, was that ". . . the killer coronavirus probably originated from a laboratory in Wuhan."[40]

Both of the Wuhan labs were known to be collecting, analyzing, and experimenting on hundreds of live bat viruses, at times making them more virulent and infective. Mind you, these labs are located in a city of 10 million people—a classic environment primed for quick spread of viruses. Doing gain-of-function research is unquestionably a primary threat to human existence—why would anyone conduct this dangerous research on virulent viruses in such a densely populated area?

A number of scientists who worked in these labs—most notably Dr. Shi Zhengli, dubbed the "Bat Woman" by the Chinese press—had published peer-reviewed articles in scientific journals, often co-authored by US and other foreign scientists, describing how they used gain-of-function techniques such as genetic engineering and lab manipulation to make coronaviruses more infectious and virulent.[41]

Later it would come out that these gain-of-function experiments were being funded and carried out, not only by the Chinese government and the military, but also by the US government, including Dr. Anthony Fauci's National Institute of Allergy and Infectious Diseases (NIAID) and the Eco-Health Alliance, among others, as well as a shadowy network of Pentagon and national security agencies.[42]

In an Instagram post Robert Kennedy, Jr., described the complicity of Dr. Anthony Fauci, the supposed "rational voice" of the Trump administration on COVID-19, in funding the weaponization of viruses in the Wuhan labs:

The Daily Mail today reports that it has uncovered documents showing that Anthony Fauci's NIAID gave $3.7 million to scientists at the Wuhan Lab at the center of Coronavirus leak scrutiny.

According to the British paper, "the federal grant funded experiments on bats from the caves where the virus is believed to have originated." Background: following the 2002–2003 SARS coronavirus outbreak, NIH funded a collaboration by Chinese scientists, US military virologists from the bioweapons lab at Ft. Detrick, and NIH scientists from NIAID to prevent future coronavirus outbreaks by studying the evolution of virulent strains from bats in human tissues.

Those efforts included "gain of function" research that used a process called "accelerated evolution" to create COVID Pandemic superbugs: enhanced bat-borne COVID mutants more lethal and more transmissible than wild COVID. Fauci's studies alarmed scientists around the globe who complained, according a Dec. 2017 NY Times article, that "these researchers risk creating a monster germ that could escape the lab and seed a pandemic."

Dr. Mark Lipsitch of the Harvard School of Public Health's Communicable Disease Center told the Times that Dr. Fauci's NIAID experiments "have given us some modest scientific knowledge and done almost nothing to improve our preparedness for pandemic, and yet risked creating an accidental pandemic." In October 2014, following a series of federal laboratory mishaps that narrowly missed releasing these deadly engineered viruses, President Obama ordered the halt to all federal funding for Fauci's dangerous experiments.

It now appears that Dr. Fauci may have dodged the federal restrictions by shifting the research to the military lab in Wuhan. Congress needs to launch an investigation of NIAID's mischief in China.[43]

As Kennedy points out, part of the reason why some of these gain-of-function experiments were being carried out in Wuhan from 2014 until the present time was because there was supposed to be a ban (from 2014 to 2017) on such experiments in the US, following a series of lab accidents and leaks in the US and abroad.[44]

Unfortunately, lab accidents, thefts, and leaks have become all too commonplace, not only in the US but in China as well, including SARS-CoV (the viral ancestor of SARS-CoV-2), which escaped from labs in Bejing, Singapore, and other locations in 2003 and 2004 and infected and even killed lab workers. As

Science magazine warned in 2004: "With four separate infections within the last year at three different institutions in Beijing, Singapore, and Taipei, health experts fear that the next SARS epidemic may be more likely to emerge from a research lab than from the presumed animal reservoir."[45]

Ominously, in January 2018, following a site visit to the Wuhan Institute of Virology, several members of the US State Department sent warnings back to Washington that the lab appeared to be dangerously mismanaged and inadequately staffed, posing a serious risk of an accidental release of a potential pandemic pathogen (PPP).[46]

Facing mounting skepticism and criticism from independent investigators, Gao Fu, director of the Chinese Center for Disease Control and Prevention, admitted in late May 2020, on Chinese national television, that no SARS-CoV-2 viruses had been detected in animal samples at the Wuhan market, and that therefore the Wuhan seafood market was not the source of the epidemic, but rather a location where previously infected people spread the virus.

"At first, we assumed the seafood market might have the virus, but we now realize that the market is more like a victim."[47] Gao's admission was consistent with multiple studies showing that the COVID-19 virus was circulating in Wuhan before any person was infected at the seafood market.[48]

Ever-Changing Twists in the Official Story

Most of the world media and health authorities, no doubt somewhat embarrassed and perplexed, gave very little coverage at first to the retraction of the wet-market theory on the part of the Chinese government.

Chinese officials and gain-of-function scientists then quickly began to change the official origin story, claiming that the coronavirus host species, horseshoe bats, must have somehow infected a lone animal in a shipment of wild pangolins, smuggled in from Malaysia, 1,000 miles from the bat caves, in a miraculous genetic recombination incident that enabled the bat coronavirus to become highly infectious and therefore capable of detonating a pandemic.

But then, once again, after a stream of articles promoting the bat-to-pangolin thesis, and cries to crack down on wildlife smuggling, scientists around the world began publishing papers pointing out that the bat-to-pangolin hypothesis was not a plausible explanation for the sudden emergence of SARS-CoV-2.[49]

As many scientists have pointed out, it is highly unlikely that the original SARS bat coronavirus (basically non-transmissible to humans) coexisted and then exchanged genes with a rare pangolin to develop a special spike protein called a furin cleavage site, making it highly transmissible to humans.

This proposed recombination event was especially unbelievable (unless it was genetically engineered in a lab) given that bats and pangolins live thousands of miles apart. In addition, if this hypothetical, highly infectious bat/pangolin virus ever did exist, it would be extremely unlikely for it to leave no biological or epidemiological traces whatsoever in its wake. In nature this type of complex cross-species recombination, experts pointed out, would take decades to evolve to the point of infecting the first humans, and then even more time to evolve further to become highly infectious to humans.

But there is absolutely no evidence that this bat/pangolin virus ever existed, or ever infected anyone, despite bat hunters, epidemiologists, and virologists all over the world searching for evidence. If a version of this bat/pangolin coronavirus did exist, it must have been created in a lab, where it escaped or was released in the late fall of 2019 in Wuhan.

Other apologists for the natural origins of SARS-CoV-2 claimed there would be traces or scars left on the genome of the SARS-CoV-2 virus where insertions had occurred, had it been genetically engineered. Still others claimed the lack of a known "viral backbone" for SARS-CoV-2 (necessary for genetic engineering techniques in a lab) invalidated the hypothesis.

But as Moreno Colaiacovo, an Italian genome data scientist, and many other scientists have pointed out, modern genetic engineering procedures can create new viral constructs while leaving absolutely no trace ("seamless techniques"). Every genetic engineer and virologist knows this nowadays, even if the mass media and the general public do not.

Moreover, while it is true that even the most skilled genetic engineers do need a preexisting viral backbone to manipulate and weaponize viruses in the lab, the universe of known viruses, such as the backbone of SARS-CoV-2, would not include those thousands of "publicly unknown" viruses that the Chinese military or others had in their possession but have chosen not to make public. Utilizing one of these "publicly unknown" viruses as a backbone, gain-of-function scientists could readily create a SARS-CoV-2 virus from a previous, but unpublished, virus and viral backbone in their possession.[50]

Origins of the Coronavirus: Conundrums

So where did the new SARS-CoV-2 come from? Where did it pick up its unique biological ability, which SARS-CoV (the original SARS coronavirus) was lacking, to efficiently unlock human cell defenses? Specifically, where did the original SARS-CoV coronavirus pick up the unique four-amino-acid segment, perfectly placed in its genome, that enabled it to use furin and other enzymes in the human body to dissolve its viral coating so that it can penetrate and infect human cells and start to reproduce?

Despite scientists all over the world carefully analyzing numerous samples of the SARS-CoV-2 virus that have infected humans, no evidence has been uncovered of the virus evolving from being less infectious to more infectious among humans over time.

The coronavirus was somehow "optimized" to infect humans from day one, as you would expect from something that was created in a lab, not a virus that was gradually circulating and spreading among humans outside of a laboratory setting.

As one group of researchers stated: "Our observations suggest that by the time SARS-CoV-2 was first detected in late 2019, it was already pre-adapted to human transmission to an extent similar to late epidemic SARS-CoV. However, no precursors or branches of evolution stemming from a less human-adapted SARS-CoV-2-like virus have been detected."[51]

Another group of scientists used a computer model to test the way the SARS-CoV-2 spike protein bound with the receptors of the cells of many species. They discovered that the spike protein bound more strongly with human ACE2 receptors than those of any other species. They wrote:

> *Notably, this approach surprisingly revealed that the binding energy between SARS-CoV-2 spike protein and ACE2 was highest for humans out of all species tested, suggesting that SARS-CoV-2 spike protein is uniquely evolved to bind and infect cells expressing human ACE2.*
>
> *This finding is particularly surprising as, typically, a virus would be expected to have highest affinity for the receptor in its original host species, e.g. bat, with a lower initial binding affinity for the receptor of any new host, e.g. humans. However, in this case, the affinity of SARS-CoV-2 is higher for humans than for the putative original host species, bats, or for any potential intermediary host species.[52]*

In other words, there is very strong scientific evidence that the SARS-CoV-2 virus was engineered in a laboratory, rather than being a virus that naturally evolved in bats and an intermediate species and then "spilled over" into humans, where over time it became more infectious.

Beyond the overwhelming scientific evidence, what about the fact that SARS-CoV-2 happened to appear first, in a billion-to-one coincidence, in the same urban neighborhood where Chinese scientists, in a joint US/China partnership, were collecting bat viruses from the wild, and then weaponizing them in several badly managed, accident-prone labs, under the guise of biomedicine and vaccine research?

This doesn't mean, of course, that SARS-CoV-2 was deliberately released, but it does point to the extremely high probability that it escaped from a lab, already primed to infect humans.

Perhaps the Chinese and American scientists who champion gain-of-function research were trying to create a more virulent and infectious coronavirus so they could infect mice or other lab animals with it and then try to develop a vaccine, rather than creating a bioweapon.

In fact, Dr. Shi Zhengli, who worked in the Wuhan lab, has mentioned the lab's use of transgenic mice in order to culture a specific affinity between viruses and human ACE2 receptors. (Angiotensin-converting enzyme 2, or ACE2, is a protein whose receptors are located on cell membranes, and is responsible for how the SARS-CoV-2 virus gains access to the cell.)

Many have struggled to understand how this affinity between the SARS-CoV-2 and human cells could be so pronounced, and why, if it's not the pangolin, we haven't been able to find the intermediate species that enabled the virus to jump from bats to humans. Dr. Zhengli told *Science* in July 2020:

> *We performed in vivo experiments in transgenic (human ACE2 expressing) mice and civets in 2018 and 2019 in the Institute's biosafety laboratory. The viruses we used were bat SARSr-CoV close to SARS-CoV. Operation of this work was undertaken strictly following the regulations on biosafety management of pathogenic microbes in laboratories in China. The results suggested that bat SARSr-CoV can directly infect civets and can also infect mice with human ACE2 receptors. Yet it showed low pathogenicity in mice and no pathogenicity in civets.*[53]

This explains why Big Pharma, or Anthony Fauci's National Institutes of Health, would fund this kind of controversial lab research in Wuhan, even after most US-based gain-of-function research was temporarily banned (between 2014 and 2017) as too dangerous.[54]

But you can be sure that the Chinese military and the US military and security agencies, which are the largest funders of this type of biowar/biodefense research across the globe, see a virus like SARS-CoV-2 as a potential bioweapon, especially in the context of a global chemical and biological arms race.[55] Unfortunately, as scientist and author André Leu, director of Regeneration International, points out, it is ". . . highly unlikely that the Wuhan Institute of Virology researchers and the Chinese government will ever tell the truth given the immense scale of the cover-up . . . They know if the truth gets out about how gain-of-function research has caused this global pandemic, that has wrecked the lives of millions, the outcry and anger will be so great that this type of research will be banned."[56]

Additionally, the Chinese and US governments, which funded these reckless experiments, and the military-industrial complex and their scientists who carried them out, would be liable for trillions of dollars in damages arising from the COVID-19 pandemic and would likely face criminal charges if it can be proven they violated the global treaty banning the development of chemical and biological weapons.

Shades of Pandemics Past

There have been several other instances of novel coronaviruses that forecast our global doom. The first was the Spanish flu of 1918, caused by an avian virus that managed to cross over to both pigs and humans. It hit during World War I in 1918, infected 500 million people worldwide, killing an estimated 50 million, or 2.7 percent of the global population.[57]

The Spanish flu was a very rapid killer, causing death in as little as 12 hours. Like the novel coronavirus SARS-CoV-2, the virus also spread very easily and rapidly. Unlike the case of COVID-19, however, people between the ages of 20 and 40 were most susceptible to the infection. With COVID-19, it's the elderly and immune-compromised who are at greatest risk, but even in these high-risk groups, the mortality rate is nowhere near that of the Spanish flu. While there is much talk about how COVID-19 is an event similar in effect to the Spanish flu, it shares far more similarities with later bird and swine flu scares.

In 1976 a novel swine flu infected 230 soldiers at Fort Dix, New Jersey, causing one death. Fearing a repeat of the 1918 Spanish flu pandemic, a vaccine was fast-tracked and the government propaganda machine cranked into action, telling all Americans to get vaccinated. What had been a contained outbreak resulted in a massive swine flu vaccine campaign in which more than 45 million Americans were vaccinated.

Over the next few years, nearly four thousand Americans filed vaccine damage claims with the federal government[58] totaling $3.2 billion.[59] Side effects included several hundred cases of Guillain-Barré syndrome (a rare side effect of flu vaccines). Even healthy 20-year-olds ended up as paraplegics. At least 300 deaths were also attributed to the vaccine.[60] Meanwhile, the death tally from this "pandemic virus" itself never rose above one.

Then there was the H5N1 bird flu scare that cropped up in 2005—the one that caused President Bush to declare that two million Americans would die as a result.[61] Those bird flu fears were exposed as little more than a cruel hoax, designed to instill fear and line the pocketbooks of various individuals and industry. Dr. Mercola's *New York Times*–bestselling book, *The Bird Flu Hoax*, details the massive fraud involved with the epidemic that never happened.

One long-lasting effect of those bird flu scares, however, is that the WHO started to coordinate a fast-track procedure for licensing and approval of pandemic vaccines. As noted on the WHO's website: "Ways were sought to shorten the time between the emergence of a pandemic virus and the availability of safe and effective vaccines."[62]

One such method used in Europe is to conduct *advance* studies using a "mock-up" vaccine that contains an active ingredient for an influenza virus that has not circulated recently in human populations. When testing these mock-up vaccines, it is very possible to release the novel influenza virus into the population, as its purpose is to "mimic the novelty of a pandemic virus" and "greatly expedite regulatory approval."

Then came the swine flu scare of 2009. In that year major news outlets warned that the swine flu could kill 90,000 Americans and hospitalize 2 million. It was an echo of the fearmongering that went on during the 2005 bird flu pandemic, which never materialized.

In response to the 2009 swine flu pandemic, what did the Centers for Disease Control and Prevention suggest? Swine flu shots for all! As the *Washington Post* reported, the CDC said: "As soon as a vaccine is available, try to get it for everyone in your family."[63] This, even though the severity of the 2009 H1N1 virus was moderate—generally requiring neither hospitalization nor even medical care. In fact, most cases had mild symptoms that cleared up on their own.

The fast-tracked 2009 swine flu vaccine for the European market (Pandemrix) turned out to be a disaster. In 2011 it was causally linked[64] to childhood narcolepsy, which had abruptly skyrocketed in several countries.[65] In 2019 researchers discovered a "novel association between Pandemrix-associated narcolepsy and the non-coding RNA gene GDNF-AS1"—a gene thought to regulate the production of glial-cell-line-derived neurotrophic factor or GDNF, a protein that plays an important role in neuronal survival. They also confirmed a strong association between vaccine-induced narcolepsy and a certain haplotype, suggesting that "variation in genes related to immunity and neuronal survival may interact to increase the susceptibility to Pandemrix-induced narcolepsy in certain individuals." [66]

As with COVID-19, there's evidence to suggest that the 2009 swine flu was the result of genetic engineering and a lab accident. A 2009 article published in the *New England Journal of Medicine* said:[67] "Careful study of the genetic origin of the virus showed that it was closely related to a 1950 strain but dissimilar to influenza A (H1N1) strains from both 1947 and 1957. This finding suggested that the 1977 outbreak strain had been preserved since 1950.[68] The reemergence was probably an accidental release from a laboratory source in the setting of waning population immunity to H1 and N1 antigens."[69]

You must ask yourself, who stands to benefit from all of this paranoia and hysteria? You undoubtedly know the answer to this one. Big Pharma, of course, but also Big Ag, Big Tech, and the technocrats who are seeking to instill a New World Order.

COVID-19: Planned or Simply Predicted and Exploited?

Dr. Anthony Fauci himself said: "And if there's one message that I want to leave with you today based on my experience . . . [it] is that there is no question that there will be a challenge to the coming [Trump] administration in the arena of infectious diseases . . . but also there will be a surprise outbreak . . . The thing we're extraordinarily confident about is that we're going to see this in the next few years."[70]

We are looking at what is possibly the greatest crime or act of criminal negligence and cover-up in modern times. In a criminal trial, if a suspect (the Chinese government), or a group of suspects (the Chinese and the American governments and their scientists), act like they are guilty, concealing or destroying evidence, intimidating witnesses, attacking their critics as "conspiracy theorists," and constantly changing their story or their alibi, then they probably are guilty, or else they are trying to cover up for someone else.

If someone, possibly an accomplice or a collaborator of these perpetrators, benefits economically or politically or in terms of increased power and control from a crime or a disaster, or from covering up the real origins of this crime or disaster, we need to look at them and their statements or testimony very carefully.

If someone, or in this case a group of very wealthy and powerful people (including Bill Gates, the World Health Organization, and the World Economic Forum), uncannily predicts in great detail that a pandemic like

COVID-19 is about to emerge (which they did in a high-level exercise known as Event 201), and then it happens, we need to sit up and take notice. Especially when this same global elite then proceeds to manipulate the disaster to their advantage, carefully controlling and manipulating the narrative and rolling out ambitious plans for what amounts to a New World Order with technocratic and totalitarian control, known as The Great Reset.[71] (We will look at Event 201 and the Great Reset in detail in the next chapter.)

The Verdict

There is a growing body of evidence, forensic and circumstantial, telling us we should be skeptical of the official story of the origins of COVID-19, put out and defended by the Chinese and US governments, Big Pharma, the scientific establishment, Big Tech, the World Economic Forum, the World Health Organization, Bill Gates, and the military-industrial complex, among others.

Weighing the scientific data and the growing preponderance of circumstantial evidence (suspects, behavior, money, motives, rewards, beneficiaries, social control, history of accidents and releases, anticipation or prediction), we inevitably reach the conclusion that SARS-CoV-2 emerged from a lab, rather than naturally, either because of an accident (which seems more likely) or because it was deliberately released.[72] But we need to know what lab, what scientists, what virus, and what viral constructs leaked out.

We need a global public inquiry led by independent scientists, lawyers, and investigators to gather the evidence on what really happened with COVID-19. This should be followed by an international biowarfare crimes tribunal, along the lines of the post–World War II Nuremberg trials, so that we can bring the perpetrators of this pandemic to justice and prevent this type of disaster from ever happening again.

Once we have identified the COVID-19 perpetrators and their accomplices, and exposed them, along with those who have funded, aided, and abetted them, the global grassroots must revisit the Biological Weapons Convention. We must move to strengthen the prohibitions and close all the loopholes in the global treaty banning chemical and biological weapons, including all weaponization of viruses, bacteria, and microorganisms.

Part of this amended treaty must include putting an end to all bioweapons research, denying militarists and genetic engineers the loophole of calling bioweapons research "biomedical" or "biosafety" research, and calling for mandatory inspections (just as are now required for nuclear weapons) whenever and wherever violations of the international treaty are suspected.

As bioweapons expert Lynn Klotz has pointed out:

> *After its enactment in 1975, one criticism of the major international treaty banning biological weapons, the Biological Weapons Convention, was that it had no provisions to monitor whether countries were complying with it. Being, as it was, the middle of the Cold War, it was unlikely that the Soviet Union would allow international inspectors to visit its biodefense facilities . . .*
>
> *After the Soviet Union collapsed, countries sought to address this perceived weakness . . . by [providing] procedures for randomly selected site visits and a rapid means to investigate weapons development, stockpiling, and use. But supporters of the proposal had their hopes dashed in 2001 when the United States pulled out of a UN ad hoc group tasked with drafting the protocol, meaning the proposed provisions never were enacted into international law.*[73]

It's time to shut down every dual-use biowar/biosafety lab in the world and implement a true global ban on weapons of mass destruction, including all chemical and biological weapons and experimentation.

Then and only then will we be able to fully understand and defend ourselves from COVID-19 and prevent the next pandemic, which Bill Gates has already warned us is coming, in the form of a bioterrorism attack.[74]

Until we do this, none of us will ever be safe again.

CHAPTER THREE

Event 201 and the Great Reset

By Dr. Joseph Mercola

While it may be hard to believe, the evidence suggests the COVID-19 pandemic is anything but accidental. As I will review in this chapter, simulations done a mere 10 weeks before the outbreak were eerily identical to the events that have played out in the real world. At the same time, technocrats around the globe were quick to use the pandemic as justification for rolling out plans that have been decades in the making behind the scenes.

While it's difficult to identify who the technocratic elite are, experts like Patrick Wood, an economist, financial analyst, and American constitutionalist who has devoted a lifetime to researching and understanding technocracy, suggest we look at private, global organizations that play a leading role in shaping our global economies and social and environmental movements.

While technocracy used to be an actual private club, the technocrats of today do not necessarily have membership cards. Key players, however, are the members of the Trilateral Commission. Well-known names in the US Trilateral group include Henry Kissinger, Michael Bloomberg, and Google heavyweights Eric Schmidt and Susan Molinari, the company's vice president for public policy. Other groups to look at include:

- The Club of Rome.
- The Aspen Institute, which has groomed and mentored executives from around the world about the subtleties of globalization. Many of its board members are also members of the Trilateral Commission.
- The Atlantic Institute.
- The Brookings Institution and other think tanks.

The World Health Organization (WHO), the medical branch of the UN, also plays a central role in the technocratic plan, as does the World Economic Forum (WEF), which serves as the social and economic branch of the UN and is the organization that hosts the annual conference of billionaires at Davos, Switzerland. The World Economic Forum was founded by Klaus Schwab, who also wrote the books *The Fourth Industrial Revolution* (2016), *Shaping the Fourth Industrial Revolution* (2018), and *COVID-19: The Great Reset.*

The Bill and Melinda Gates Foundation became the WHO's largest funder when the US government, in mid-April 2020, halted funding until a White House review of the WHO's handling of the COVID-19 pandemic could be completed.[1] Gavi, the Vaccine Alliance, a partnership between Gates and Big Pharma with a stated aim of solving global health problems through vaccines, is also a top donor to the WHO and one of the primary initiatives of the WEF.[2] The way Klaus Schwab describes Gavi says a great deal: "In many ways [Gavi] is a role model for how the public and private sector can and should cooperate—working in a much more efficient way than governments alone or business alone or civil society alone."[3]

It may sound appealing, until you realize that they are working efficiently to strip us of our liberties.

The World Economic Forum is a conglomeration of the world's largest and most powerful businesses, all of which are helping to further the technocratic agenda along. They include Microsoft, which made Bill Gates a billionaire; MasterCard, which is leading the globalist charge to develop digital IDs and banking services; Google, the number-one Big Data collector in the world and a leader in AI services; as well as foundations started by the world's wealthiest people, such as the Rockefeller Foundation, the Rockefeller Brothers Fund, the Ford Foundation, Bloomberg Philanthropies, and George Soros's Open Society Foundations.[4]

When you peek behind the curtain at the WEF and the WHO, you find all the same wealthy individuals and their companies and foundations who, although they claim to be working for a more equitable society and healthier planet, are really only trying to centralize profit and power.

Many of the terms we've heard more and more of in recent years also refer to technocracy under a different name. Examples include sustainable development, Agenda 21, the 2030 Agenda, the New Urban Agenda, green economy, the green new deal, and the global warming movement in general. They all refer to and are part of technocracy and resource-based economics. Other terms that are synonymous with technocracy include the Great Reset,[5] the Fourth Industrial Revolution,[6] and the slogan Build Back Better.[7] The Paris Climate Agreement is also part and parcel of the technocratic agenda.

The common goal of the Great Reset and of all these movements and agendas is to capture all of the resources of the world—the ownership of them—for a small global elite group that has the know-how to program the computer systems that will ultimately dictate the lives of everyone. It's really the ultimate form of totalitarianism. When they talk about "wealth redistribution," what they're actually referring to is the redistribution of resources from us to them. A glimpse into this future was offered in a November 2016 *Forbes* article written by Ida Auken from the World Economic Forum leadership strategy team. It reads, in part:

> *Welcome to the year 2030. Welcome to my city—or should I say, "our city." I don't own anything. I don't own a car. I don't own a house. I don't own any appliances or any clothes . . . Everything you considered a product, has now become a service . . . In our city we don't pay any rent, because someone else is using our free space whenever we do not need it. My living room is used for business meetings when I am not there . . . Once in a while I get annoyed about the fact that I have no real privacy. Nowhere I can go and not be registered. I know that, somewhere, everything I do, think and dream of is recorded. I just hope that nobody will use it against me. All in all, it is a good life.*[8]

If you rent everything and have no private property of your own, then who does own all of those things? The technocratic elite who own all the energy resources. Disturbingly enough, one form of energy resource that modern technocrats apparently intend to harvest, if patents are any indication, is the human body. As just one example, Microsoft's international patent WO/2020/060606 describes a "cryptocurrency system using body activity data."[9] This patent, if implemented, would essentially turn human beings into robots. People will be brought down to the level of mindless drones, spending their days carrying out tasks automatically handed out by, say, a cellphone app, in return for a cryptocurrency "award."

The Public Face of Technocracy: Bill Gates

Once you become familiar with the technocratic agenda, you can start to recognize the players rather easily. One of the most obvious ones is Bill Gates. Almost everything he does furthers the technocratic agenda.

Gates, who cofounded Microsoft in 1975, is perhaps one of the most dangerous philanthropists in modern history, having poured billions of dollars into global health initiatives that stand on shaky scientific and moral ground—including the COVID-19 pandemic.

Gates's answers to the problems of the world are consistently focused on building corporate profits through highly toxic methods, be they chemical agriculture and GMOs, or pharmaceutical drugs and vaccines. Rarely, if ever, do we find Gates promoting clean living or inexpensive holistic health strategies, and we've certainly seen that during this pandemic. Vaccines and various surveillance technologies have been his go-to answers throughout, and these are the very industries he has vested interests in.

Gates Donates Billions to Private Companies

A March 17, 2020, article in *The Nation* titled "Bill Gates' Charity Paradox" details "the moral hazards surrounding the Gates Foundation's $50 billion charitable enterprise, whose sprawling activities over the last two decades have been subject to remarkably little government oversight or public scrutiny."

As noted in this article, Gates discovered an easy way to gain political power —"one that allows unelected billionaires to shape public policy"—namely charity. Gates has described his charity strategy as "catalytic philanthropy," in which the "tools of capitalism" are leveraged to benefit the poor.

The only problem is that the true beneficiaries of Gates's philanthropic endeavors tend to be those who are already rich beyond comprehension, including Gates's own charitable foundation. The poor, on the other hand, end up with costly solutions like patented GMO seeds and vaccines that in some instances have done far more harm than good.

In addition to donations given to nonprofit organizations, Gates also donates to for-profit, private companies. According to *The Nation*, the Gates Foundation has given close to $250 million in charitable grants to companies in which the foundation holds corporate stocks and bonds.[10] In other words, the Gates Foundation is giving money to companies from which it will benefit financially in return for its "donations." As a result, the more money Gates and his foundation give, the more their wealth grows. Part of this growth in wealth also appears to be due to the tax breaks given for charitable donations. In short, it's a perfect money-shuffling scheme that allows him to evade taxes while maximizing income generation.

Gates's "philanthropy" has certainly played a central role in the COVID-19 pandemic, and here, too, he is benefiting handsomely—again, by investing in the industries he's giving charitable donations to, and by promoting a global public health agenda that benefits the companies he's invested in.

Virtually every aspect of the pandemic involves organizations, groups, and individuals funded by Gates. This includes the World Health Organization, of course, but also the two research groups responsible for shaping the decisions to lock down the U.K. and US—the Imperial College COVID-19 Research Team and the Institute for Health Metrics and Evaluation.

Neil Ferguson, a professor of mathematical biology at Imperial College London, has produced a string of pandemic predictions that have turned out to be spectacularly incorrect, including his 2005 forecast that 200 million people would die from bird flu.[11] Meanwhile, in the real world, the final death toll ended up being just 282, worldwide, between 2003 and 2009.[12]

In 2020 Ferguson's Imperial College model for COVID-19—relied on by governments around the world—led to the most draconian pandemic response measures in modern history.[13] It predicted the U.K. would be looking at a death toll of more than 500,000, and the US some 2.2 million, if no action was taken. This is precisely the kind of convenient disinformation and gross overestimation of risk that Gates needs and relies on to drive his own vaccine and tech agendas forward.

That Gates's philanthropic endeavors protect his own investments can also be seen in his pro-patent stance. James Love, director of the nonprofit Knowledge Ecology International, pointed out to *The Nation* that Gates ". . . uses his philanthropy to advance a pro-patent agenda on pharmaceutical drugs, even in countries that are really poor . . . He's undermining a lot of things that are really necessary to make drugs affordable . . . He gives so much money to [fight] poverty, and yet he's the biggest obstacle on a lot of reforms."[14]

Gates is a staunch and longtime defender of the drug industry, and his intent to further the pharmaceutical agenda can clearly be seen in the current COVID-19 pandemic. From the very beginning, Gates was out in front saying that nothing will go back to normal until or unless the entire global population gets vaccinated and countries implement tracking and tracing technologies and "vaccine passports." At the same time, he's pouring money into digital ID projects and cashless society plans. Ultimately, all of these things will be connected, forming a "digital prison" in which the technocratic elite will have complete control over the global population.

Buying Favorable Press

While Gates has faced public backlash a number of times in his career, especially when he was CEO of Microsoft in the 1990s, he's become increasingly insulated from negative reviews, thanks to the fact that he also funds journalism and major media corporations.

In an August 21, 2020, article in *Columbia Journalism Review*, Tim Schwab highlights the connections between the Bill and Melinda Gates Foundation and a number of newsrooms, including NPR. These outlets routinely publish news favorable to Gates and the projects he funds and supports. Not surprisingly, experts quoted in such stories are almost always connected to the Gates Foundation as well.

Schwab examined the recipients of nearly 20,000 Gates Foundation grants, finding that more than $250 million had been given to major media companies, including BBC, NBC, Al Jazeera, ProPublica, *National Journal*, the *Guardian*, Univision, Medium, the *Financial Times*, *The Atlantic*, the *Texas Tribune*, Gannett, *Washington Monthly*, *Le Monde*, PBS NewsHour, and the Center for Investigative Reporting. (The time frame of those grants is unfortunately unclear.)

The Gates Foundation has also given grants to charitable organizations that in turn are affiliated with news outlets, such as BBC Media Action and the *New York Times*'s Neediest Cases Fund.

Journalistic organizations such as the Pulitzer Center on Crisis Reporting, the National Press Foundation, the International Center for Journalists, the Solutions Journalism Network, and the Poynter Institute for Media Studies have also received grants from the Gates Foundation.

Ironically, "The foundation even helped fund a 2016 report from the American Press Institute that was used to develop guidelines on how newsrooms can maintain editorial independence from philanthropic funders," Schwab writes.

The Gates Foundation has also participated in dozens of media conferences, including the Perugia Journalism Festival, the Global Editors Summit and the World Conference of Science Journalists, and has an unknown number of undisclosed contracts with media companies to produce sponsored content.

Upon scrutiny, it becomes abundantly obvious that when Gates hands out grants to journalism, it's not an unconditional handout with which these companies can do whatever they see fit. It comes with significant strings, and really amounts to little more than the purchasing of stealth self-promotions that are essentially undisclosed ads.

Another recipient of grants from the Gates Foundation is the Leo Burnett Company, an advertising agency that creates news content and works with journalists, and which you will see come into further play later in this chapter.

Event 201—Dress Rehearsal for COVID-19

There's a lot of evidence pointing to COVID-19 being a planned event that is now being milked for all it's worth, even though it didn't turn out to be nearly as lethal as initially predicted. In October 2019, just 10 weeks before the COVID-19 outbreak first began in Wuhan, China, the Bill and Melinda Gates Foundation co-hosted a pandemic preparedness simulation of a "novel coronavirus," known as Event 201, along with the Johns Hopkins Center for Health Security and the World Economic Forum.

This scripted tabletop included everything that has since played out in the real world, from PPE shortages, lockdowns, censorship, and removal of civil liberties to mandated vaccination campaigns, riots, economic turmoil, and the breakdown of social cohesion.

Just as in real life, "misinformation" they said would need to be countered included rumors that the virus had been created and released from a bioweapons laboratory and questions surrounding the safety of fast-tracked vaccines.

Johns Hopkins University may seem like a reputable institution, but consider that it was started by the Rockefeller Foundation, and that researchers from the Rockefeller Foundation and Johns Hopkins University were behind the infamous and cruel experiments on 600 Black sharecroppers in Tuskegee, Alabama—who were injected by researchers with syphilis without their consent and then were never given actual treatment, only placebos, even as they infected their wives and children.

Rockefeller Foundation and Johns Hopkins researchers were also involved in the horrific Guatemala experiments that occurred between 1946 and 1948, when 5,000 vulnerable Guatemalans, including prostitutes, orphans, and the mentally ill, were barbarically infected with bacteria containing multiple sexually transmitted diseases, including syphilis and gonorrhea.[15]

Bradley Stoner, MD, past president of the American Sexually Transmitted Diseases Association, described the Guatemala experiments as "something right out of Dr. Mengele's notebook"—a reference to the experiments Jews endured at the hands of the Nazis during the Second World War.[16] Together, the Gates Foundation, the World Economic Forum, and Johns Hopkins University form what appears to be a technocratic triad, whose pandemic simulation was more of a dress rehearsal than anything.

Event 201 Predicted "Need" for Censorship

Event 201 planners spent a great deal of time discussing ways to limit and counter the spread of expected "misinformation" about the pandemic and subsequent vaccines. In addition to outright censoring certain views, Event 201 introduced a plan that included the use of "soft power," a term referring to stealth influencing. This strategy uses celebrities and other social media influencers to model ideal behavior and promote adherence to pandemic response edicts.

Take for example Tom Hanks and his wife, Rita Wilson, both of whom reportedly tested positive for COVID-19 early on in the pandemic. They dutifully modeled the desired behavior—getting tested, self-quarantining, and submitting to continued observation for as long as necessary to ensure they didn't spread it to anyone else—and shared their every step on social media and in traditional media outlets. That's one example of soft power.

Celebrities also put on a virtual "One World Together at Home" benefit concert to raise money for the WHO and rally citizens of the world around the idea that we can get through this if we all just follow instructions and stay home. In May 2020 celebrities and social media influencers agreed to "pass the mic" by allowing the WHO and other pandemic response leaders, such as Dr. Anthony Fauci, to use their social media accounts to share their messages.

If you thought all of these things occurred more or less organically, you'd be wrong. *Daily Caller* spilled the beans in the July 17, 2020, article "World Health Organization Hired PR Firm to Identify Celebrity 'Influencers' to Amplify Virus Messaging."[17] According to *Daily Caller*:

> *The World Health Organization hired a high-powered public relations firm to seek out so-called influencers to help build trust in the organization's coronavirus response.*
>
> *WHO paid $135,000 to the firm Hill and Knowlton Strategies, according to documents filed under the Foreign Agents Registration Act. . . . The contract earmarked $30,000 for "influencer identification," $65,000 for "message testing," and $40,000 for a "campaign plan framework."*[18]
>
> *Hill and Knowlton . . . proposed identifying three tiers of influencers: celebrities with large social media followings, individuals with smaller but more engaged followings, and "hidden heroes," those users with slight followings but who "nevertheless shape and guide conversations."*

The WHO isn't the only organization trying to control the narrative, of course. Many other organizations are involved, all working toward the same end. The United Nations, for example, enlisted 10,000 "digital volunteers" to rid the internet of what they consider "false" information about COVID-19 and to disseminate what they say is "U.N.-verified, science-based content."

The campaign, dubbed the Verified initiative,[19] amounts to an army of internet trolls engaging in censorship in an attempt to shut down opposition and opinions that run counter to the status quo.

Who's in Charge of Determining What Is True?

The UN's Verified campaign is reminiscent of another self-appointed internet watchdog, NewsGuard, which claims to rate information as "reliable" or "fake" news, supplying you with an authoritative color-coded rating badge next to Google and Bing searches, as well as on articles displayed on social media.

If you rely on NewsGuard's ratings, you may decide to entirely skip articles from sources with a low red rating in favor of the so-called more trustworthy green-rated articles—and therein lies the problem. NewsGuard is in itself fraught with conflict of interest, as it's largely funded by Publicis, a global communications giant that's partnered with Big Pharma and WEF, such that it may be viewed more as a censorship tool than an internet watchdog.[20]

What's more, the Leo Burnett Company, also owned by Publicis, is a recipient of grants from none other than the Gates Foundation. On top of that, News-Guard and Microsoft—the tech company founded by Gates—are also partners.[21]

For example, NewsGuard announced that Mercola.com has been classified as fake news because we have reported the SARS-CoV-2 virus as potentially having been leaked from the biosafety level 4 (BSL 4) laboratory in Wuhan City, China, the epicenter of the COVID-19 outbreak. NewsGuard's position is in direct conflict with published scientific evidence suggesting that this virus was created in a lab and not zoonotically transmitted.

Using the Pandemic to Further the Tyrannical Loss of Liberty

Can you start to see the picture of a larger agenda forming?

For decades, the threat of conflict and the fear of attacks have provided the justification needed for war and military occupations as well as the chipping away of our civil liberties. The Patriot Act, rammed through in the aftermath of 9/11, is just one egregious example.

The hysteria whipped up around 9/11 and the anthrax attacks created the conditions for the passage of the Patriot Act—a 342-page document that was clearly already written, not composed in just two weeks after the attack[22]— which changed 15 existing laws and allowed the TSA to legally record anyone's phone calls. This was all under the guise of protecting "freedom," when in reality it was one of the biggest steps toward the loss of civil liberties in the history of the US.

The Patriot Act was rushed through Congress while, ominously, two con-gressmen who opposed it—Senator Tom Daschle from South Dakota and Senator Patrick Leahy from Vermont—had letters with weaponized, military-grade anthrax mailed to their offices.

We can now point to the passage of the Patriot Act as the technocratic elite's first step toward taking away many of our constitutional rights and personal freedoms and laying down the foundation for a modern surveillance/ police state. From the ACLU:

> *Hastily passed 45 days after 9/11 in the name of national security, the Patriot Act was the first of many changes to surveillance laws that made it easier for the government to spy on ordinary Americans by expanding the authority to monitor phone and email communications, collect bank and credit reporting records, and track the activity of innocent Americans on the Internet. While most Americans think it was created to catch terrorists, the Patriot Act actually turns regular citizens into suspects.*[23]

In short, the Patriot Act normalized invasive surveillance and the removal of privacy rights. Today, pandemics and the threat of infectious outbreaks and bioterrorism are the new tools of war and social control. For the authors of this book, the manipulations and fearmongering that pave the way for a surveillance state are far more dangerous and insidious than the viral infection itself. The global technocratic elite are making George Orwell's book *1984* a reality. Between the Patriot Act and pandemic measures, the groundwork has been laid for the Great Reset.

For over a decade, Gates also prepared the global psyche for a new enemy: deadly, invisible viruses that can crop up at any time.[24] And according to Gates, the only way to protect ourselves is by giving up "old-fashioned" notions of privacy, liberty, and personal decision making.

Thanks to the COVID-19 pandemic, we need to maintain our physical distance from others, including family members. We are told to wear masks, even in our own homes and during sex. Small businesses have been forced to close, many of which have gone bankrupt as a result. Office workers are told to work from home. We're told we have to vaccinate the entire global population and enforce stringent travel restrictions to prevent spread. (See chapter 8 for information on the coronavirus vaccine.) We are being tracked and traced every moment of the day and night, and there are plans to implant biometric readers into everyone's bodies to identify who the potential risk-carriers are. Infected people are the new threat, the new invisible enemy. This is what the technocratic elite (see the "Technocracy Defined" sidebar, page 45, for a breakdown of this term), spearheaded by Gates, want you to believe, and it's really surprising how, in a matter of months, they've been able to convince most of the population of this.

But the technocratic agenda currently playing out was most certainly in place long before the pandemic began. In 2017, Gavi, the Vaccine Alliance, which was founded by the Gates Foundation in partnership with the WHO, the World Bank, and various vaccine manufacturers, decided to provide every child with a digital biometric identity that would store his or her vaccination records.

Shortly thereafter, Gavi became a founding member of the ID2020 Alliance, alongside Microsoft and the Rockefeller Foundation. In 2019 Gates collaborated with Massachusetts Institute of Technology professor Robert Langer to develop a novel vaccine delivery method using fluorescent microdot tags—essentially creating an invisible "tattoo" that can then be read with a modified smartphone.

As investigative journalist James Corbett points out in his *Corbett Report* segment titled "Bill Gates and the Population Control Grid":[25]

> *It should be no surprise, then, that Big Pharma vaccine manufacturers—in their scramble to produce the coronavirus vaccine that, Gates assures us, is necessary to "go back to normal"—have turned to a novel vaccine delivery method: a dissolvable microneedle array patch . . . As in so many other aspects of the unfolding crisis, Gates' unscientific pronouncement that we will need digital certificates to prove our immunity in the "new normal" of the post-coronavirus world is now being implemented by a number of governments.*

In his coverage of Bill Gates, Corbett also reviews the rapid development and implementation of biometric identification programs tied in with digital currencies. Undoubtedly, the plan is to connect everything together—your identification, your personal finances, and your medical and vaccination records. Most likely, it will also be embedded on your body, for your own "convenience," so you cannot lose it. Never mind the fact that everything that can be hacked at some point has been or will be. On top of that, Western nations can expect the rollout of a social credit system similar to that in China. In December 2020 the International Monetary Fund presented a plan to tie people's credit scores to their internet search histories.[26]

As noted by Corbett:

> *The ID control grid is an essential part of the digitization of the economy . . . And although this is being sold as an opportunity for "financial inclusion" of the world's poorest in the banking system provided by the likes of Gates and his banking and business associates, it is in fact a system for financial exclusion. Exclusion of any person or transaction that does not have the approval of the government or the payment providers . . .*

The different parts of this population control grid fit together like pieces of a jigsaw puzzle. The vaccination drive ties into the biometric identity drive which ties into the cashless society drive.

In Gates' vision, everyone will receive the government-mandated vaccinations, and everyone will have their biometric details recorded in nationally administered, globally integrated digital IDs. These digital identities will be tied to all of our actions and transactions, and, if and when they are deemed illegal, they will simply be shut off by the government—or even the payment providers themselves.

Indeed, if you think online censorship is bad, consider a world in which your online activity is tied to your biometric chip with all your finances and personal data. What easier way to silence people than to block access to their own money? We're sure there are many other ways in which such a system could be used to control any and all individuals.

Corbett continues:

Only the most willfully obtuse could claim to be unable to see the nightmarish implications for this type of all-seeing, all-pervasive society, where every transaction and every movement of every citizen is monitored, analyzed, and databased in real-time by the government . . . And Bill Gates is one of those willfully obtuse people. This Gates-driven agenda is not about money. It is about control. Control over every aspect of our daily lives, from where we go, to who we meet, to what we buy and what we do.

Facebook: A Tool for Social Control

The backbone and infrastructure of technocracy is technology. Stunning capacities to surveil, analyze, and manipulate our behavior already exist—and the power of technology is advancing at an exponential rate.

Back in March 2020, when the pandemic first began, the White House Office of Science and Technology Policy began assembling a task force of tech and artificial intelligence companies to "develop new text and data-mining techniques that could help the science community answer high-priority scientific questions related to COVID-19," according to CNBC.[27] Not surprisingly, the 60 companies included Facebook, which currently creates and shares "disease-prevention maps" derived from aggregated user data with the government, researchers, and nonprofits.

Technocracy Defined

This chapter is about the technocratic elite and their technocratic agenda, which is being pushed through via manipulation of the pandemic. But what exactly is technocracy? The work of Patrick Wood has helped us as we've sought to understand the foundational cause of the problem at hand.

If you are interested in taking a deeper dive into technocracy, we recommend his books *Technocracy Rising: The Trojan Horse of Global Transformation* and *Technocracy: The Hard Road to World Order*.

In summary, technocracy is a resource-based economic system that began in the 1930s during the height of the Great Depression, when scientists and engineers got together to solve the nation's economic problems. It looked like capitalism and free enterprise were failing, so they decided to invent a new economic system from scratch. They called this system technocracy.

Rather than being based on pricing mechanisms such as supply and demand, technocracy is based on resource allocation and social engineering through technology. Under this system, companies would be told what resources they're allowed to use, when, and for what, and consumers would be told what to buy.

Artificial intelligence (AI), digital surveillance, and Big Data collection play very important roles, as does the digitization of industries and government, such as banking and health care. Together these technologies allow for the automation of social engineering and social rule, thereby doing away with the need for elected government leaders. Nations are to be led by unelected leaders who own all the world's resources and decide what is to become of them.

Technocrats have silently and relentlessly pushed this agenda forward for decades, and it's now becoming increasingly visible, with world leaders openly calling for a global "reset" of the economy and how we live in general.

The only reason technocracy has not yet been able to completely overtake the US—although, as you're now seeing, they're getting incredibly close—is because of the US Constitution. This is why we must fight to protect our Constitution at all costs, through grassroots movements and getting involved in local politics.

When people use Facebook apps on their phones, maps are generated, though the information is not shared with the general public.[28] Facebook says the maps, generated by a project called Data for Good:[29]

> . . . are designed to help public health organizations close gaps in understanding where people live, how people are moving, and the state of their cellular connectivity, in order to improve the effectiveness of health campaigns and epidemic response.
>
> These datasets, when combined with epidemiological information from health systems, assist nonprofits in reaching vulnerable communities more effectively and in better understanding the pathways of disease outbreaks that are spread by human-to-human contact.

Despite assurances of anonymity and no plan to track individuals, enlisting Big Tech companies to work directly with the government is concerning when it comes to preserving your privacy. Who can forget the 2018 scandal in which Cambridge Analytica, a political data firm, gained access to private information on more than 50 million Facebook users?[30]

Though Facebook says its data is anonymized, only shows general trends, and is not used to track individuals, the task force plans would enlarge Facebook's role in providing data to the government. Facebook CEO Mark Zuckerberg said privacy concerns around tracking fears are "overblown." Moreover, while some tech companies already share aggregated data generated by users, *Wired* notes that ". . . it would be new for Google and Facebook to openly mine user movements on this scale for the government. The data collected would show patterns of user movements. It would need to be cross-referenced with data on testing and diagnoses to show how behavior is affecting the spread of the virus."

Caroline Buckee, associate professor at the Harvard TH Chan School of Public Health, told *Wired* that though aggregated, anonymized location data is already available from Google, Facebook, Uber, and phone companies, the worry is that the collected data will be reverse-engineered to track people.

Privacy suspicions do not just stem from the Cambridge Analytica scandal. During the Washington State COVID-19 outbreak, Facebook data were fed into models produced by the Institute for Disease Modeling in Bellevue, which collaborates with, shocker, the Bill and Melinda Gates Foundation and other groups.

Forbes reported that Gates called for a "national tracking system similar to South Korea . . . to understand where the disease is and whether we need to strengthen the social distancing" in response to the COVID-19 epidemic.[31]

Gates responded to a question during a Reddit "Ask Me Anything" session by saying: "Eventually we will have some digital certificates to show who has recovered, or been tested recently, or, when we have a vaccine who has received it."[32]

"Digital certificates" . . . are you seeing the puzzle and picture coming more and more together now? You can be assured that virtually everything you do and say online is mined and manipulated by the social media companies.

We operate under their control; we separate into tribes, fight with one another, and live in fear, and this is a highly effective way to ensure control. Social media, tracking devices, 5G, satellites, artificial intelligence . . . even though it sounds like a dystopian science-fiction novel, it's becoming painfully obvious that we are far along in following the plots of futuristic movies like *Terminator* and *The Matrix*. We're watching it happen in real time.

The Great Reset

By now, you've probably started hearing world leaders speak of "the Great Reset," "the Fourth Industrial Revolution," and the call to "Build Back Better." As mentioned earlier, all of these terms refer to the new social contract planned for the world, which is just a new term for the New World Order.

"The Great Reset" was introduced in mid-2020 by the World Economic Forum. Yes, *that* World Economic Forum—the organization that partnered with Gates to host Event 201.

The leaders of the WEF, the WHO, the UN, and their partner organizations have had this idea for a long, long time. A conglomeration of the world's largest and most powerful businesses has been working toward the Great Reset, which ultimately boils down to the greatest wealth transfer in the history of the world. It's a long-term plan to disempower and disenfranchise all but the wealthiest by monitoring and controlling the world through technical surveillance. While a world war would have been ideal, President Trump's peace efforts appear to have put a damper on that strategy, resulting in the pandemic being used as the justification for a reset instead.

As the WEF clearly points out, after the Great Reset is implemented, you won't own anything. What they don't tell you is that the partners at the WEF will own everything instead, and that your willingness to follow their rules will be directly tied to how many provisions you are allotted.[33]

Ultimately, the technocratic agenda seeks to integrate humankind into a technological surveillance apparatus overseen by powerful artificial intelligence. Ironically, while the real plan is to usher in a tech-driven dystopia free of democratic controls, they speak of this plan as a way to bring us back into harmony with nature.

According to the World Economic Forum, the Great Reset "will address the need for a more fair, sustainable and resilient future, and a new social contract centered on human dignity, social justice and where societal progress does not fall behind economic development."[34]

They're using feel-good terms like *sustainability, social justice, food justice, climate-smart agriculture*, and *poverty reduction*. And that's on purpose: They know people want these things, so they're saying that's what their plan offers. The price, however, is your personal liberty. In his report, investigative journalist James Corbett summarizes the Great Reset thus: "At base, the Great Reset is nothing more, and nothing less, than a great propaganda, marketing rollout campaign for a new brand that the would-be global elite are trying to shove down the public's throats . . . It's just a fresh coat of lipstick on a very old pig. This is The New World Order, just redefined. It's just a new label for it."

And as explained by Corbett, for those who forgot about what the New World Order was all about, it was all about "centralization of control into fewer hands, globalization [and] transformation of society through Orwellian surveillance technologies." In other words, it's technocracy, where we the people know nothing about the ruling elite while every aspect of our lives is surveilled, tracked, and manipulated for their gain. Far from being the end of globalization, the Great Reset is globalization turbocharged. The plan is not to "reset" the world back to some earlier state that will allow us all to start over with a cleaner environment and more equitable social structures. No, the plan is to circumvent democracy and shift global governance into the hands of the few.

As noted by Klaus Schwab in his book, *COVID-19: The Great Reset*:[35]

> *When confronted with it, some industry leaders and senior executives may be tempted to equate reset with restart, hoping to go back to the old normal and restore what worked in the past: traditions, tested procedures and familiar ways of doing things—in short, a return to business as usual. This won't happen because it can't happen. For the most part, "business as usual" died from (or at the very least was infected by) COVID-19.*

Build Back Better

To be sure, the pandemic has caused widespread economic devastation. So don't we need to "build back better?" Make no mistake, this catchy slogan is part and parcel of the Great Reset plan and cannot be separated from it, no matter how altruistic it may sound. Joe Biden, whose campaign slogan in his winning 2020 presidential bid was "Build Back Better," has a long record of being anti-privacy and pro-technology.

According to a 2008 CNET article:

> *On privacy, Biden's record is hardly stellar. In the 1990s, Biden was chairman of the Judiciary Committee and introduced a bill called the Comprehensive Counter-Terrorism Act . . . A second Biden bill was called the Violent Crime Control Act. Both were staunchly anti-encryption, with this identical language:*
>
> *"It is the sense of Congress that providers of electronic communications services and manufacturers of electronic communications service equipment shall ensure that communications systems permit the government to obtain the plain text contents of voice, data, and other communications when appropriately authorized by law." Translated, that means turn over your encryption keys.*[36]

But the phrase *build back better* did not originate with Biden. In fact, it was introduced by the UN in a press release. It reads:

> *As the world begins planning for a post-pandemic recovery, the United Nations is calling on Governments to seize the opportunity to 'build back better' by creating more sustainable, resilient and inclusive societies. "The current crisis is an unprecedented wake-up call," said Secretary-General António Guterres in his International Mother Earth Day message. "We need to turn the recovery into a real opportunity to do things right for the future."*[37]

The UN has directly exhorted nations worldwide to "build back better" after Covid-19,[38] and the phrase has been widely adopted by government leaders in Great Britain,[39] New Zealand,[40] Canada, and elsewhere. In addition to decimating privacy, part of the "building back better" plan is to shift the financial system over to a central bank digital currency (CBDC) system,[41] which in turn is part of the system of social control, as it can easily be used to incentivize desired behaviors and discourage undesired ones.

There's general agreement among experts that most major countries will implement CBDC within the next two to four years. Contrary to popular belief, these CBDCs will not be anything like existing cryptocurrencies like

Bitcoin. While Bitcoin is decentralized and a rational strategy to opt out of the existing central-bank-controlled system, CBDCs will be centralized and completely controlled by the central banks, and will have smart contracts that allow the banks to surveil and control your life.

Be Afraid, Be Very Afraid

It goes without saying that to achieve this kind of radical transformation of every part of society has its challenges. No person in their right mind would agree to it if they were aware of the details of the whole plan. So to roll out the Great Reset, the elite have had to use psychological manipulation, and fear is the most effective tool there is.

As explained by psychiatrist Dr. Peter Breggin, there's an entire school of public health research that focuses on identifying the most effective ways to frighten people into accepting desired public health measures.

By adding confusion and uncertainty to the mix, you can bring an individual from fear to anxiety—a state of confusion in which you can no longer think logically—and in this state you are more easily manipulated. Figure 3.1 illustrates the central role of fearmongering for the successful rollout of the Great Reset.

Keep in mind that, as we've illustrated in this chapter, technocracy is inherently a technological society run through social engineering. Fear is but one manipulation tool. The focus on "science" is another. Anytime someone dissents, they're simply accused of being "anti-science," and any science that conflicts with the status quo is declared "debunked science."

The only science that matters is whatever the technocrats deem to be true, no matter how much evidence there is against it. We've seen this firsthand during this pandemic, as Big Tech has censored and banned anything going against the opinions of the World Health Organization, which is just another cog in the technocratic machine.

If we allow this censorship to continue, our civil liberties will be rapidly eroded and replaced with tyrannical suppression of the constitutional rights that our ancestors fought and died for. We simply must keep pushing for transparency and truth. We must insist on medical freedom, personal liberty, and the right to privacy.

One fight in particular that I don't see us being able to evade is the fight against mandatory COVID-19 vaccinations. If we don't take a firm stand against that and fight for the right to make our own choice, there will be no end to the medical tyranny that will follow. We will cover vaccines much more

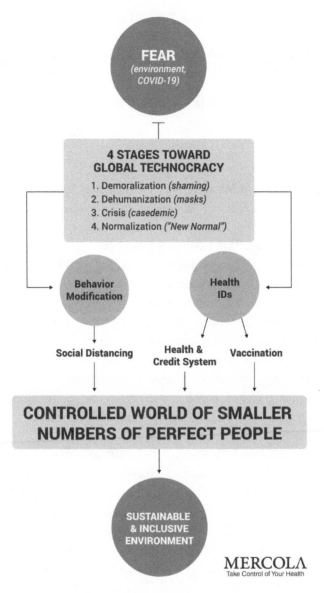

Figure 3.1. Technocracy and the Great Reset: Psychological Operations Guide.

in chapter 8. Meanwhile, let's turn to examining the virus itself more closely to evaluate its true dangers—or lack thereof—so that we can start to dissipate some of the fear that the technocracy is counting on to make their Great Reset easier to implement.

CHAPTER FOUR

COVID-19 Strikes the Most Vulnerable

By Dr. Joseph Mercola

You know the official story: COVID-19 is a highly contagious and deadly infection that can be stopped only by social distancing, frequent handwashing, lockdowns, masks, mass testing, contact tracing, and ultimately vaccines. But in reality, COVID-19 appears to be a highly contagious, dangerous, lab-manufactured "trigger" for the preexisting conditions of an aging and increasingly chronically ill population.

The virus itself isn't the primary cause of most COVID-19 hospitalizations and fatalities. Rather, the virus exploits other serious diseases with high mortality that are widespread in the population and dangerous in and of themselves. It's these comorbidities, along with rampant medical malpractice (and other factors we've already touched on and will cover further in this book), that are the main drivers of COVID-19 hospitalizations and deaths. To put it simply: People are dying *with* COVID-19 as opposed to dying *from* it.

Data Show COVID-19 Isn't a Significant Threat

To understand the truth versus the official story, we have to separate the real statistics from the "official" statistics on cases, hospitalizations, and deaths. A relatively high "case" load does not mean people are actually getting sick and dying. The media has been conflating a positive test result with the actual disease, COVID-19, thereby deliberately misleading the public into believing the infection is far more serious and widespread than it actually is.

COVID-19 is *not* confirmed by a positive test; it is a clinical diagnosis of someone infected with SARS-COV-2 who is exhibiting severe respiratory illness characterized by fever, coughing, and shortness of breath. By using a test that falsely labels healthy individuals as sick and infectious, mass testing drives the narrative that we're in a lethal pandemic.

Indeed, the use of reverse transcription polymerase chain reaction (RT-PCR) tests is at the very heart of this entire scam. If it wasn't for this flawed test, there would be no pandemic to speak of. I will review this in greater detail in chapter 5.

Mislabeled Causes of Death

According to groundbreaking data released by the CDC on August 26, 2020, only *6 percent* of the total COVID-19-related deaths in the US had COVID-19 listed as the sole cause of death on the death certificate.[1]

To help that sink in: 6 percent of 496,112 (the total death toll reported by the CDC as of February, 21, 2021) is 29,766. In other words, SARS-CoV-2 infection was directly responsible for 29,766 deaths of otherwise healthy individuals—a far different story from the 200,000-plus (and rising) number reported in the media. The remaining 94 percent of patients had an average of 2.6 health conditions that contributed to their deaths.

These data paint a picture that's in stark contrast with Johns Hopkins University, which in August 2020 reported that about 170,000 of the 5.4 million Americans who had tested positive for COVID-19 had died, prompting Dr. Thomas Frieden, former director of the US Centers for Disease Control and Prevention, to say that COVID-19 is now the third leading cause of death in the US, killing more Americans than "accidents, injuries, lung disease, diabetes, Alzheimer's, and many, many other causes."[2] Frieden is simply stoking the flames of fear with this claim.

Johns Hopkins has been having a hard time keeping its story straight. In November 2020 the institution published an article alleging accounting errors on a national level regarding COVID-19 deaths in the elderly.

"Surprisingly, the deaths of older people stayed the same before and after COVID-19," the author of the article said. "Since COVID-19 mainly affects the elderly, experts expected an increase in the percentage of deaths in older age groups. However, this increase is not seen from the CDC data. In fact, the percentages of deaths among all age groups remain relatively the same."

But after a link to the Johns Hopkins article was posted on Twitter, the article quickly disappeared.[3] Fortunately, an archive of it is still available.[4]

The American Institute for Economic Research reported on the mysterious disappearance of the article and went a few steps further by posting its own graph taken from CDC data in April 2020. "This suggests that it could be possible that a large number of deaths could have been mainly due to more serious ailments such as heart disease but categorized as a COVID-19 death, a far less lethal disease," the institute reported.[5] Incidentally, this is precisely what CDC guidance has instructed medical practitioners to do.

The CDC's Plan to Intentionally Inflate Numbers of Deaths Due to COVID-19

The CDC has done its part to ensure that as many deaths as possible are attributed to COVID-19—even when it was not the actual cause of death. In personal correspondence, Meryl Nass, MD, reported that in March 2020: "The CDC issued new guidance that required doctors who complete death certificates to list COVID-19 on the certificate if it contributed to or caused the death. This was no different than what we did before. We are supposed to list all contributory causes."

The official communication at that time read:

> It is important to emphasize that Coronavirus Disease 2019 or COVID-19 should be reported on the death certificate for all decedents where the disease caused or is assumed to have caused or contributed to death . . .
>
> For example, in cases when COVID-19 causes pneumonia and fatal respiratory distress, both pneumonia and respiratory distress should be included along with COVID-19 in Part I . . . If the decedent had other chronic conditions such as COPD or asthma that may have also contributed, these conditions can be reported in Part II.[6]

In April 2020 the CDC issued new guidance documents on how to complete death certificates for COVID-19[7] and even hosted a webinar on the process, but according to Nass, the guidelines remained substantively the same. Then, later in the fall of 2020, the CDC changed course dramatically, this time *without* bringing any attention to the new guidelines. According to Nass: "Without fanfare, the CDC acknowledged on another webpage that even if COVID was *not* listed by the doctor as the underlying cause of death, or the proximate cause of death, as long as it was listed as one cause or contributor, it would be coded as *the* cause of death."

Indeed, the CDC website at the time of this writing reads (emphasis ours): "When COVID-19 is reported as *a* cause of death on the death certificate, it is coded and counted as a death *due to* COVID-19."[8]

All of this caused Nass to conclude that the fanfare that occurred in April was "deliberate misdirection." You may not appreciate how absurd this is, so let me give you an example. If a young healthy person died in a motorcycle accident and had tested positive for SARS-CoV-2, according to these CDC guidelines, their death would be listed as a COVID-19 death.

All these machinations with the death certificates hide the fact that the death rate from COVID-19 for everyone except for those over 60 is significantly lower than the death rate for influenza.

COVID Versus Influenza

Though an article in *Scientific American* called the claim that the virus's fatality is on par with the flu "fake news,"[9] there's nothing fake about it. We call your attention to research looking at the fatality ratio for the average person, excluding those residing in nursing homes and other long-term care facilities, presented September 2, 2020, in *Annals of Internal Medicine*: "The overall non-institutionalized infection fatality ratio [for COVID-19] was 0.26 percent . . . Persons younger than 40 years had an infection fatality ratio of 0.01 percent; those aged 60 or older had an infection fatality ratio of 1.71 percent."[10]

Other sources are reporting similar findings. During an August 16, 2020, lecture at the Doctors for Disaster Preparedness convention, Dr. Lee Merritt pointed out that, based on deaths per capita—which is the only way to get a true sense of the lethality of this disease—the death rate for COVID-19 at that time was around 0.009 percent.[11] That number was based on a global total death toll of 709,000, and a global population of 7.8 billion. This also means the average person's chance of surviving an encounter with SARS-CoV-2 was 99.991 percent.

In comparison, the estimated infection fatality rate for seasonal influenza listed in the *Annals of Internal Medicine* paper is 0.8 percent. Other sources put it a little higher. In either case, *the only people for whom SARS-CoV-2 infection is more dangerous than influenza are those over the age of 60*. All others have a lower risk of dying from COVID-19 than they have of dying from the flu.

White House coronavirus task force coordinator Dr. Deborah Birx also confirmed this far lower than typically reported mortality rate when she, in mid-August 2020, stated that it "becomes more and more difficult" to get people to comply with mask rules "when people start to realize that 99 percent of us are going to be fine."[12]

Who Gets Sick?

In April 2020 nearly all crew members of the deployed aircraft carrier USS *Theodore Roosevelt* were tested for SARS-CoV-2. By the end of the month, of the roughly 4,800 crew on board, 840 tested positive. However, 60 percent were asymptomatic, meaning they had no symptoms. Only one crew member died, and none were in intensive care.[13]

Similarly, among the 3,711 passengers and crew aboard the *Diamond Princess* cruise ship, 712 (19.2 percent) tested positive for SARS-CoV-2, and of these 46.5 percent were asymptomatic at the time of testing. Of those showing symptoms, only 9.7 percent required intensive care and 1.3 percent died.[14]

Military personnel, as you would expect, tend to be healthier than the general population. Still, the data from these two incidents reveal several important points to consider. First of all, it suggests that even when living in close, crowded quarters, the infection rate is rather low.

Only 17.5 percent of the USS *Theodore Roosevelt* crew got infected—slightly lower than the 19.2 percent of those aboard the *Diamond Princess*, which had a greater ratio of older people.

Second, fit and healthy individuals are more likely to be asymptomatic than not—60 percent of naval personnel compared with 46.5 percent of civilians onboard the *Diamond Princess* had no symptoms despite testing positive.

Medical Errors Responsible for Most COVID-19 Deaths

Now that we've established that the official statistics aren't telling us the whole truth and that COVID-19 isn't responsible for nearly as many deaths as we've been told, let's look at a leading cause of death that you *don't* hear about in the media: medical malpractice.

In 2016 a Johns Hopkins study found that more than 250,000 Americans die each year from preventable medical errors, effectively making modern medicine the third leading cause of death in the US.[15] Other estimates place the death toll from medical mistakes as high as 440,000.[16] The reason for the discrepancy in the numbers is that medical errors are rarely noted on death certificates, and death certificates are what the CDC relies on to compile its death statistics.

While medical errors are continually swept under the proverbial rug, they need to be brought to light now more than ever, because they play also play a role in the death toll attributed to COVID-19.

A significant portion of those who have died from COVID-19 were in fact victims of medical errors. In particular, Elmhurst Hospital Center in Queens, New York—which was "the epicenter of the epicenter" of the COVID-19 pandemic in the US—appears to have grossly mistreated COVID-19 patients, thereby causing their death.[17]

Financial Incentives Increased Deaths

According to army-trained nurse Erin Olszewski, who worked at Elmhurst during the height of the outbreak in New York City, hospital administrators and doctors made a long list of errors, most egregious of which was to place all COVID-19 patients, including those merely suspected of having COVID-19, on mechanical ventilation rather than less invasive oxygen administration.

During her time there, most patients who entered the hospital wound up being treated for COVID-19, whether they tested positive or not, and only one patient survived. The hospital also failed to segregate COVID-positive and COVID-negative patients, thereby ensuring maximum spread of the disease among non-infected patients coming in with other health problems.

By ventilating COVID-19-negative patients, the hospital artificially inflated the caseload and death rate. Disturbingly, financial incentives appear to have been at play. According to Olszewski, the hospital received $29,000 extra for a COVID-19 patient receiving ventilation, over and above other reimbursements. In August 2020, CDC director Robert Redfield admitted that hospital incentives likely elevated hospitalization rates and death toll statistics around the country.[18]

Many Governors Radically Increased Elderly Deaths with Misguided Policies

Another major error that drove up the death toll was state leadership's decision to place infected patients in nursing homes, against federal guidelines.[19] According to an analysis by the Foundation for Research on Equal Opportunity, which included data reported by May 22, 2020, an average of 42 percent of all COVID-19 deaths in the US had occurred in nursing homes, assisted living facilities, and other long-term care facilities.[20]

This is extraordinary, considering this group accounts for just 0.62 percent of the population. By and large nursing homes are ill equipped to care for COVID-19-infected patients.[21] While they're set up to care for elderly patients—whether they are generally healthy or have chronic health problems—these facilities are rarely equipped to quarantine and care for people with highly infectious diseases.

It's logical to assume that commingling infected patients with non-infected ones in a nursing home would result in exaggerated death rates, as the elderly are far more prone to die from any infection, including the common cold. We also learned, early on, that the elderly were disproportionately vulnerable to severe SARS-CoV-2 infection.

Yet ordering infected patients into nursing homes with the most vulnerable population of all is exactly what several governors decided to do, including New York's Andrew Cuomo, Pennsylvania's Tom Wolf, New Jersey's Phil Murphy, Michigan's Gretchen Whitmer, and California's Gavin Newsom.[22]

ProPublica published an investigation on June 16, 2020, comparing a New York nursing home that followed Cuomo's misguided order with one that refused, opting to follow the federal guidelines instead. The difference was stark.[23]

By June 18 the Diamond Hill nursing home—which followed Cuomo's directive—had lost 18 residents to COVID-19, thanks to lack of isolation and inadequate infection control. Half the staff (about 50 people) and 58 patients were infected and fell ill.

In comparison, Van Rensselaer Manor, a 320-bed nursing home located in the same county as Diamond Hill, which refused to follow the state's directive and did not admit any patient suspected of having COVID-19, did not have a single COVID-19 death. A similar trend has been observed in other areas.

Ventilators Did Not Help and Only Increased Deaths

The misuse of mechanical ventilation was not limited to Elmhurst Hospital Center in Queens. As early as June 2020, researchers warned that COVID-19 patients placed on ventilators are at increased risk of death, and leading experts suggested the machines were being overused and that patients would likely do better with less invasive treatments. According to one study, more than 50 percent of mechanically ventilated COVID-19 patients died.[24]

The practice remained widespread, nonetheless. In a case series of 1,300 critically ill patients admitted to intensive care units (ICUs) in Lombardy, Italy, 88 percent received invasive ventilation, but the mortality rate was still 26 percent.[25] Further, in a *JAMA* study that included 5,700 patients hospitalized with COVID-19 in the New York City area between March 1, 2020, and April 4, 2020, mortality rates for those who received mechanical ventilation ranged from 76.4 percent to 97.2 percent, depending on age.[26]

Similarly, in a study of 24 COVID-19 patients admitted to Seattle-area ICUs, 75 percent received mechanical ventilation and, overall, half of the patients died between 1 and 18 days after being admitted.[27]

There are many reasons why those on ventilators have a high risk of mortality, including being more severely ill to begin with. There are risks inherent to mechanical ventilation itself, including lung damage caused by the high pressure used by the machines. In cases of acute respiratory distress syndrome (ARDS), the lung's air sacs may be filled with a yellow fluid that has a "gummy" texture, making oxygen transfer from the lungs to the blood difficult, even with mechanical ventilation. Long-term sedation from the intubation is another significant risk that is difficult for some patients, especially the elderly, to bounce back from.

A Perfect Storm of Errors

Novel viruses always have their highest impact at the beginning of their existence before their impact levels off. A never-before-seen virus is like touching a spark to dry wood. It burns hottest in the beginning, before fairly quickly cooling down.

With a novel virus, the most vulnerable are hit rapidly. In the case of the SARS-CoV-2, nursing homes were the dry wood. Due to the combination of the vulnerable being hit first and the medical community mistreating those who became ill, the initial spike in fatalities was real, although it didn't have to be as high as it was.

If it weren't for systematic medical mistreatment at certain hospitals, widespread erroneous use of ventilators, and incomprehensible decision making by a handful of state governors, the COVID-19 death toll may well have been negligible.

When you add all of these factors together—the wanton mismanagement of the infection in hot spots such as New York, the decision to send infected patients into nursing homes, the fact that few healthy people died from the infection, plus that potential medical treatments have been and still are actively suppressed—it very much appears to be a manufactured crisis.

Sepsis May Be at the Root of Many COVID-19 and Influenza Deaths

Sepsis is a life-threatening condition triggered by a systemic infection that causes your body to overreact and launch an excessive and highly damaging immune response. A number of studies have shown that sepsis is becoming ever more prevalent. In the US, 1.7 million adults develop sepsis each year, and nearly 270,000 die as a result.[28] In fact, between 34.7 percent and 55.9 percent of American patients who died in hospitals between 2010 and 2012 had sepsis at the time of their death.[29]

Worldwide, sepsis is responsible for one in five deaths each year—double the rate of previous estimates—according to the most comprehensive global analysis to date. The researchers call the finding "alarming." As reported by NPR: "They estimate that about 11 million people worldwide died with sepsis in 2017 alone—out of 56 million total deaths. That's about 20 percent of all deaths."[30]

A significant hurdle when studying sepsis is the fact that many doctors overlook it as a contributing cause of death and don't list it on the death certificate. Yet sepsis has been identified as a major contributor in influenza deaths.

One of the problems is that the symptoms of sepsis are easy to confuse with a bad cold, influenza, and COVID-19—including dehydration, high fever, chills, confusion, rapid heartbeat, nausea or vomiting, and cold, clammy skin. However, they tend to develop more quickly than you would normally expect. Unless promptly diagnosed and treated, sepsis can rapidly progress to multiple-organ failure and death.

Severe sepsis is traditionally associated with bacterial diseases. However, viruses are becoming a growing cause of severe sepsis worldwide—including COVID-19. In fact, in July 2020 famous Broadway actor Nick Cordero died of complications from COVID-19, including septic shock, or sepsis. Cordero is by no means the only one. Sepsis is an important contributor to the death of many COVID-19 patients—one that's been flying largely under the radar.

According to Dr. Karin Molander, chair of the Sepsis Alliance board of directors, "sepsis is a leading, if not the number one, fatal complication of COVID-19."[31] Sepsis occurs so often alongside COVID-19 that the National Center for Health Statistics released updated guidelines for medical coding of the two conditions.[32]

Many Critically Ill COVID Patients Develop Viral Sepsis

Researchers from China wrote in *The Lancet*: "In clinical practice, we noticed that many severe or critically ill COVID-19 patients developed typical clinical manifestations of shock, including cold extremities and weak peripheral pulses, even in the absence of overt hypotension. Understanding the mechanism of viral sepsis in COVID-19 is warranted for exploring better clinical care for these patients."[33]

Viral sepsis can be particularly challenging, according to the Sepsis Alliance, because tests that reveal bacterial sepsis to physicians do not necessarily reveal viral sepsis. That being said, abnormal vital signs, including blood pressure, pulse and respirations, may occur with either bacterial or viral sepsis.

According to Sepsis Alliance, "the elderly, very young and people with chronic illnesses or weakened immune systems" are most at risk of sepsis. While those affected often have underlying health conditions, even healthy people can be affected. "[W]hen a healthy person becomes severely ill with sepsis, it could be that their healthy immune system was so strong it triggered a cytokine storm," the Sepsis Alliance explained.[34]

Cytokines are a group of proteins that your body uses to control inflammation. If you have an infection, your body will release cytokines to help combat inflammation, but sometimes it releases more than it should. If the cytokine release spirals out of control, the resulting "cytokine storm" becomes dangerous and is closely tied to sepsis.

A sepsis treatment protocol developed by Dr. Paul Marik, which involves intravenous vitamin C with hydrocortisone and thiamine (vitamin B_1), has been shown to dramatically improve chances of survival in sepsis cases. If you suspect that you or a loved one may have sepsis, visit mercola.com and search for the article titled "Vitamin C, B_1 and Hydrocortisone Dramatically Reduce Mortality from Sepsis." It could save your or their life.

Comorbidities Are the Primary Cause of COVID-19 Hospitalizations and Deaths

To be fair, the official story and statistics have reported that underlying health conditions such as obesity, heart disease, and diabetes *are* key factors in COVID-19 fatalities. Yet the data show they're more than contributing factors: They're the *primary drivers* of hospitalizations and deaths.

In one study more than 99 percent of people who died from COVID-19-related complications had underlying medical conditions. Among those fatalities, 76.1 percent had high blood pressure, 35.5 percent had diabetes, and 33 percent had heart disease.[35]

Another study revealed that among 18- to 49-year-olds hospitalized due to COVID-19, obesity was the most prevalent underlying condition, just ahead of hypertension.[36] What's more, investigations reveal that most COVID-19 patients have more than one underlying health issue. A study looking at 5,700 New York City patients found that 88 percent had more than one comorbidity. Only 6.3 percent had just one underlying health condition, and 6.1 percent had none.[37]

Most chronic conditions—particularly diabetes and high blood pressure— have roots in metabolic dysfunction, as people with metabolic dysfunction have compromised immune systems. For detailed information on correcting metabolic dysfunction, refer to my previous bestselling book, *Fat for Fuel*.

Let's look at some of these co-factors more in-depth.

Metabolic Health

The common thread connecting nearly all of the COVID-19 comorbidities is insulin resistance. Insulin resistance is largely related to the transition to industrially processed foods and a reliance on carbohydrates over healthy fats. However, likely the most serious contributor is an increase in a specific omega-6 polyunsaturated fatty acid called linoleic acid (LA).

This fat is present in vegetable oils, which are more accurately known as seed oils. They did not exist 150 years ago, so our consumption used to be zero. Today it has increased to an average of about 80 grams a day. Excessive LA is far more dangerous than eating excessive sugar, as these fats destroy your metabolic machinery and stay in your body for years.

LA is highly perishable and prone to oxidation. As the fat oxidizes, it breaks down into harmful subcomponents, which is how LA contributes to the massive increase in heart disease, cancer, diabetes, obesity, and age-related blindness. They create inflammation and damage important tissues, especially your mitochondria, which are responsible for generating most of the energy in your body by converting your food and combining it with oxygen to create ATP.

When you have high levels of LA, your mitochondria become damaged and crippled and can't provide your body with enough fuel to repair the damage from all the inflammation and oxidative stress. This leads to insulin resistance and the development of all the comorbidities we see in COVID-19. We review the health impacts of LA further in chapter 6.

High Blood Pressure

Doctors in China quickly realized that nearly half of those dying from COVID-19 also had high blood pressure, or hypertension. Researchers used retrospective data from a hospital dedicated only to the treatment of the infection in Wuhan, China, to evaluate the association.[38]

After analyzing data from 2,877 patients, 29.5 percent of whom had a history of high blood pressure, they found that those with high blood pressure were twice as likely to die from COVID-19 than those who didn't.

Certain Drugs May Impact COVID-19 Outcomes

Making matters worse, the drugs routinely used to treat lifestyle-induced afflictions such as high blood pressure, as well as diabetes and heart disease, may also be contributing to adverse outcomes in patients with COVID-19. According to Reuters:

> A disproportionate number of patients hospitalized by COVID-19 . . . have high blood pressure. Theories about why the condition makes them more vulnerable . . . have sparked a fierce debate among scientists over the impact of widely prescribed blood-pressure drugs.
>
> Researchers agree that the life-saving drugs affect the same pathways that the novel coronavirus takes to enter the lungs and heart. They differ on whether those drugs open the door to the virus or protect against it . . . The drugs are known as ACE inhibitors and ARBs . . . In a recent interview with a medical journal, Anthony Fauci—the

US government's top infectious disease expert—cited a report showing similarly high rates of hypertension among COVID-19 patients who died in Italy and suggested the medicines, rather than the underlying condition, may act as an accelerant for the virus . . .

There is evidence that the drugs may increase the presence of an enzyme—ACE2—that produces hormones that lower blood pressure by widening blood vessels. That's normally a good thing. But the coronavirus also targets ACE2 and has developed spikes that can latch on to the enzyme and penetrate cells . . . So more enzymes provide more targets for the virus, potentially increasing the chance of infection or making it more severe.

Other evidence, however, suggests the infection's interference with ACE2 may lead to higher levels of a hormone that causes inflammation, which can result in acute respiratory distress syndrome, a dangerous build-up of fluid in the lungs. In that case, ARBs may be beneficial because they block some of the hormone's damaging effects.[39]

This presents significant challenges for patients and doctors alike, as there's currently no significant consensus on whether patients should discontinue the drugs. The Centre for Evidence-Based Medicine at the University of Oxford in England recommends switching to alternative blood pressure medicines in patients who have only mildly elevated blood pressure and are at high risk for COVID-19.

A paper in *NEJM* stressed the potential benefits of the drugs instead, saying that patients should continue taking them. However, several of the scientists who wrote that paper have done "extensive, industry-supported research on antihypertensive drugs," Reuters notes.[40] Dr. Kevin Kavanagh, founder of the patient advocacy group Health Watch USA, believes it would be unwise to allow scientists funded by the drug industry to give clinical directions at this time. "Let others without a conflict of interest try to make a call," he says.

Interestingly, while some studies have found an increased risk of COVID-19 mortality in diabetics who take statin drugs, other studies have found a protective effect. Whether statins raise the risk of mortality in severe COVID-19 or not, they do not protect you against cardiovascular disease as intended and as Big Pharma wants you to believe, and they do increase your risk of other negative health conditions. Since there are strategies you can use at home to reduce your risk of severe disease and protect your health, it is typically unnecessary and likely dangerous to seek out statin drugs. (More to come in chapter 6.)

Diabetes

When insulin resistance becomes sufficiently severe and chronic, type 2 diabetes sets in, so it's not surprising that diabetes is among the comorbidities of COVID-19. In the U.K. researchers gathered data from the National Health Service England in an effort to characterize the features of those at greatest risk of severe COVID-19.[41] The data showed that the median age of individuals hospitalized for COVID-19 was 72 years, with a hospital stay of about seven days. The most common comorbidities were chronic heart disease, diabetes, and chronic pulmonary disease.

Thus far, it's been unclear as to whether people with diabetes are more likely to get infected, but what is clear is that a disproportionate number of people with diabetes are hospitalized with severe illness. It's been estimated that 6 percent of the U.K. population has diabetes,[42] but data from the NHS England showed that 19 percent of those hospitalized had diabetes, more than three times the number in the general population.[43]

It's also important to note that while people with type 2 diabetes have double the risk of dying from COVID-19, people with type 1 diabetes are 3.5 times more likely to die from the virus than people without diabetes.[44]

In another study of 174 patients, scientists found that those with diabetes had a higher risk of severe pneumonia, excessive uncontrolled inflammation, and dysregulation of glucose metabolism.[45] Their data supports the idea that those with diabetes may experience a rapid progression of COVID-19 and that they have a poorer prognosis.

Obesity

Being obese or overweight can also raise your risk of COVID complications and death. Research suggests that even mild obesity can impact COVID-19 severity.

This finding was revealed by researchers from the Alma Mater Studiorum University of Bologna in Italy, who analyzed 482 COVID-19 patients hospitalized between March 1 and April 20, 2020. "Obesity is a strong, independent risk factor for respiratory failure, admission to the ICU and death among COVID-19 patients," they wrote, and the extent of risk was tied to a person's level of obesity.[46]

The researchers used body mass index (BMI) to define obesity in the study, finding increased risk started at a BMI of 30, or "mild" obesity. "Health care practitioners should be aware that people with any grade of obesity, not just the severely obese, are a population at risk," lead study author Dr. Matteo Rottoli said in a news release. "Extra caution should be used for hospitalized

COVID-19 patients with obesity, as they are likely to experience a quick deterioration towards respiratory failure, and to require intensive care admission."[47]

Specifically, patients with mild obesity had a 2.5 times greater risk of respiratory failure and a 5 times greater risk of being admitted to an ICU compared with non-obese patients. Those with a BMI of 35 and over were also 12 times more likely to die from COVID-19.

Similarly, a July 2020 report by Public Health England, which describes the results of two systematic reviews, found that excess weight worsened COVID-19 severity, and that obese patients were more likely to die from the disease than non-obese patients.[48]

Compared with healthy-weight patients, patients with a BMI above 25 kg/m^2 were 3.68 times more likely to die, 6.98 times more likely to need respiratory support, and 2.03 times more likely to suffer critical illness. The report also highlights data showing that the risk of hospitalization, intensive care treatment, and death progressively increases as your BMI goes up.

Age and Inflammation

All of the conditions covered thus far can cause chronic, uncontrolled inflammation, which can increase your chances of experiencing a cytokine storm. This inflammation is often called inflammaging or the "chronic low-grade inflammation occurring in the absence of overt infection." This type of damaging inflammation negatively impacts immunity.[49]

Chronic inflammation may help explain why age is such a factor in COVID-19 hospitalizations and deaths. Underlying or baseline inflammation can exacerbate the aging process and raise the risk of severe infectious disease, as has been demonstrated by the numbers of people 65 and older who have died from COVID-19. According to the Centers for Disease Control and Prevention, 8 of every 10 deaths from COVID-19 are people age 65 and older.[50]

Topping the list of factors that make the elderly more susceptible to dying is an aging immune system—both the innate and the adaptive arms. As noted by researchers Amber Mueller, Maeve McNamara, and David Sinclair: "For the immune system to effectively suppress then eliminate SARS-CoV-2, it must perform four main tasks: 1) recognize, 2) alert, 3) destroy, and 4) clear. Each of these mechanisms are known to be dysfunctional and increasingly heterogeneous in older people."[51]

During aging, your immune system undergoes a gradual decline in function known as immunosenescence, which inhibits your body's ability to recognize, alert, and clear pathogens; inflammaging is a result of this process. According to the researchers:

> *An abundance of recent data describing the pathology and molecular changes in COVID-19 patients points to both immunosenescence and inflammaging as major drivers of the high mortality rates in older patients.*
>
> *The inability of AMs [alveolar macrophages] in older individuals to recognize viral particles and convert to a pro-inflammatory state likely accelerates COVID-19 in its early stages, whereas in its advanced stages, AMs are likely to be responsible for the excessive lung damage.*

On top of the cytokine storm, perhaps what is even more predictive of death is an increase in the fibrin degradation product D-dimer that is released from blood clots in the microvasculature and is highly predictive of disseminated intravascular coagulation (DIC). The elderly have naturally higher levels of D-dimer, which appears to be a "key indicator for the severity of late-stage COVID-19," the researchers state.[52]

In the elderly, elevated levels are thought to be due to higher basal levels of vascular inflammation associated with cardiovascular disease, and this, the authors say, "could predispose patients to severe COVID-19." Similarly, the elderly tend to have higher levels of NLRP3 inflammasomes, which appear to be a key culprit involved in cytokine storms.

In chapter 6 we'll cover how we became so vulnerable in the first place. Because in order to change the future, you have to understand the past.

CHAPTER FIVE

Exploiting Fear to Lock Down Freedom

By Ronnie Cummins

The only thing we have to fear is fear itself.
—Franklin D. Roosevelt

Fear is ultimately what strips us of our human rights and drives a society into totalitarianism, and the only way to circumvent such a fate is to bravely resist fear. Today one of the biggest sources of fear is a global pandemic—one that allegedly came about naturally, and to which we have no known defenses—or so the official story goes.

Fear is one of the most potent catalysts for human behavior and we now have something no previous tyrant has had, namely the technology to track, trace, control, and manipulate individuals wherever they are. Most people are surrounded by electronics and wireless devices that harvest every imaginable data point about their personal life. That data collection is then integrated with AI-driven deep learning systems, which allows the technocratic elite to determine how to most effectively manipulate the masses.

However, as outlined in chapter 3, there is an ever-growing body of evidence that has enabled critics to dismember and discredit the "official story" on the origins, nature, dangers, prevention, and treatment of COVID-19.

This evidence clearly shows that COVID-19 and the ensuing pandemic are *not* from a previously existing relatively harmless bat coronavirus with limited transmissibility that somehow mutated so it could infect humans. Rather, it is much more likely that SARS-CoV-2 is the product of a disastrous, but unfortunately predictable, lab accident in Wuhan, China, in late 2019.

This weaponized virus, SARS-CoV-2, a joint Chinese/US creation, is likely a genetically engineered, mutant offspring of a decades-long biological arms race, disguised as gain-of-function biomedical, vaccine, or biosafety research.

For years the powers that be reassured us that genetically engineering viruses and bacteria in what are essentially unregulated bioweapons labs is safe; that the possibility of accidents, thefts, and releases of these potential pandemic pathogens

(PPPs) is vanishingly small, and therefore well worth the risk. They lied, and now we must deal with the catastrophic consequences of their criminal negligence.[1]

Lockdowns Are the Cause of Much of the COVID Damage

Did you ever wonder why the media won't name the lockdowns as the culprit of much of the damage caused by the pandemic? It's not just denialism. The official narrative is that we had no choice but to shatter life as we know it and shut everything down. Sadly, nothing could be further from the truth. No intervention like this has ever taken place in history. The lockdowns are an egregious attack on fundamental rights, liberties, and the rule of law. And the results are all around us.

Even after a full year of lockdowns, the public remains mostly deeply ignorant of the age/health gradient of COVID-19 fatalities, even though the data have been available since February 2020. According to the CDC—even conceding the inaccuracy of testing and exigencies of fatality classification—the survivability rate is 99.997 percent for 0–19 years, 99.98 percent for 20–49 years, 99.5 percent for 50–69 years, and 94.6 percent for 70-plus years.[2]

Nursing homes and hospitals have been the main vectors for disease, not social gatherings or outdoor events. The threat to school-aged kids approaches zero. The more information we get, the more normal the SARS-CoV-2 pathogen seems. It's a respiratory and flu-like illness that became pandemic before becoming endemic, just like so many other respiratory viruses over the last hundred years. We didn't shut down society, and, for that reason, we managed them just fine.

Many of us spend a good part of our day poring over the latest research, which reveals the terrible toll of the lockdowns. The inescapable horror is that this is a direct result of the *lockdowns*, not the pandemic. There's no evidence that lockdowns have actually saved lives. On the contrary, evidence shows a significant number of excess deaths are due not to COVID-19 but to drug overdoses, depression, and suicide.

The evidence also highlights the role of polymerase chain reaction (PCR) testing in driving the pandemic narrative, the falsehood of "asymptomatic transmission," the incredible proliferation of disease misclassification, and the absurdity of the idea that political solutions can intimidate and arrest a virus.

The Lockdowns Caused Massive Wealth Shift

Besides exposing the reckless gain-of-function lab origins of the virus and taking action to make sure this never happens again, we desperately need to expose the shoddy science, inaccurate lab tests, misleading statistics, and panic-mongering driving the official story on the nature and virulence of

COVID-19 and the disastrous, authoritarian measures—beneficial to the rich, disastrous to the working class, minority communities, and youth—that most governments have implemented to, supposedly, contain the virus.

Thus far, the pandemic has triggered or contributed to disease and death among the elderly and those with serious preexisting medical conditions, or comorbidities, as covered in chapter 4. It has also triggered widespread panic and fear in the general population, on a scale not seen since World War II. Panic-mongering has enabled opportunistic politicians, out-of-control scientists and genetic engineers, public health bureaucrats, and large corporations, especially Big Pharma and the tech giants, to consolidate their wealth and power as never before.

The fact that the pandemic has been used to shift wealth from the poor and middle class to the ultra-wealthy is clear for anyone to see at this point. In December 2020 the total wealth of US billionaires reached $4 trillion, more than $1 trillion of which was gained since March 2020 when the pandemic began, according to a study by the Institute for Policy Studies.[3]

While 45.5 million Americans filed for unemployment, 29 new billionaires were created, the Institute for Policy Studies reported in June 2020, and five of the richest men in the US—Jeff Bezos, Bill Gates, Mark Zuckerberg, Warren Buffett, and Larry Ellison—grew their wealth by a total of $101.7 billion (26 percent) between March 18 and June 17, 2020, alone.[4]

The reason the wealthy have only gotten richer during this pandemic is that their businesses weren't shut down. The shutdowns primarily affected small, privately owned businesses. The disparity in treatment of big-box stores and small retailers has been strikingly illogical. How is it safe to shop with hundreds of people in a Walmart but unsafe to shop in a store that can only hold a fraction of that?

Pandemic profiteers include online retailers and Big Tech companies like Amazon, Zoom, Skype, Netflix, Google, and Facebook, along with some of the largest retailers. Walmart and Target, for example, reported record sales in 2020.[5] As noted by IPS News: "The COVID pandemic has not been the 'Great Equalizer' as suggested by the likes of New York Governor Andrew Cuomo and members of the World Economic Forum. Rather, it has exacerbated existing inequalities along gender, race and economic class divides across the world."[6]

As the World Economic Forum states, "With some 2.6 billion people around the world in some kind of lockdown, we are conducting arguably the largest psychological experiment ever."[7] Our would-be global overlords openly admit that they are laying down the foundations for what they euphemistically call a Great Reset or a Fourth Industrial Revolution—a technocratic dictatorship, based upon digital surveillance, social control, and artificial intelligence, more akin to George Orwell's dystopian novel *1984* than anything else.

As a direct result of disastrous government responses, medical malpractice, and mass media panic-mongering surrounding COVID-19, the world has been turned upside down. Lockdowns, censorship, shoddy science, misleading statistics, half-truths, and outright lies have exacerbated any damage caused by the virus itself.

While the billionaire class has prospered, the global grassroots, especially the underclass, racial minorities, and children, suffer the brunt of the crisis: economic meltdown, mass unemployment, hunger, the collapse of small businesses, school closures, mass anxiety, social isolation, and unprecedented political polarization.

Back in August 2020, *Bloomberg* reported that more than half of all small business owners feared their businesses wouldn't survive.[8] They were right. According to a September 2020 economic impact report[9] by Yelp, 163,735 US businesses had closed their doors as of August 31, 2020, and of those, 60 percent—a total of 97,966 businesses—were permanent closures.[10] These business closures disproportionally affected minorities. By the end of April 2020, pandemic measures had eliminated nearly half of all Black-owned small businesses in the US.[11] According to a New York Fed report, "Black-owned businesses were more than twice as likely to shutter as their white counterparts."[12]

The Hidden Cost of Lockdowns

With unemployment comes food insecurity, and mere weeks into the pandemic, people around the world were lining up at food banks. An April 10, 2020, report by the *Financial Times* cited survey results showing an estimated three million Britons had gone without food at some point in the previous three weeks. An estimated one million people had by then already lost all sources of income.[13]

The United Nations estimates pandemic responses have "pushed an additional 150 million children into multidimensional poverty—deprived of education, health, housing, nutrition, sanitation or water,"[14] and at the end of April 2020 warned the world was facing "famine of biblical proportions," with only a limited amount of time to act before starvation claims hundreds of millions of lives.[15]

That lockdowns will have a detrimental effect on mental health also should come as no surprise, and data show that's exactly what has happened. A Canadian survey in early October 2020 found that 22 percent of Canadians experienced high anxiety levels—four times higher than the pre-pandemic rate—and 13 percent reported severe depression.[16]

In the US an August 2020 survey by the American Psychological Association found that Gen Z'ers are among the hardest hit in this regard, with young adults aged 18 to 23 reporting the highest levels of stress and depression.[17]

More than 7 out of 10 in this age group reported symptoms of depression in the two weeks before the survey. Among teens aged 13 to 17, 51 percent said the pandemic makes it impossible to plan for the future. Sixty-seven percent of college-aged respondents echoed this concern.

With despair come drug-related problems, and according to the American Medical Association, the drug overdose epidemic has significantly worsened and become more complicated this year. "More than 40 states have reported increases in opioid-related mortality as well as ongoing concerns for those with a mental illness or substance use disorder," the AMA reported in a December 9, 2020, Issue Brief.[18]

A list of national news included in the American Medical Association's brief include reports of increases in overdose-related cardiac arrests, surges in street fentanyl leading to deaths in the thousands and a "dramatic increase" in illicit opioid fatalities. Spikes and record numbers of overdose deaths have been reported in Alabama, Arizona, Arkansas, California, Colorado, Delaware, District of Columbia, Illinois, Florida and many other states.

That the lockdowns are doing more harm than good can also be seen in Centers for Disease Control and Prevention data showing that, compared with previous years, excess deaths among 25- to 44-year-olds have increased by a remarkable 26.5 percent, even though this age group accounts for fewer than 3 percent of COVID-19-related deaths.[19] To put it bluntly, in our misguided efforts to prevent the elderly and immune-compromised from dying from COVID-19, we're sacrificing people who are in the prime of their lives.

Statistics also reveal that the lockdowns have resulted in dramatic increases in domestic abuse, rape, child sex abuse, and suicides. By July 2020, Ireland reported a 98 percent increase in people seeking counseling for rape and child sex abuse.[20]

Data from the British group Women's Aid showed 61 percent of domestic abuse victims reported that their abuse had worsened during the lockdown.[21] The number of women killed by their domestic partners also doubled during the first three weeks of lockdowns in the U.K.[22]

In the US, data from a Massachusetts hospital revealed that domestic abuse cases nearly doubled in the nine weeks between March 11 and May 3, 2020, when the state had ordered schools closed.[23] Similarly, in early April 2020, United Nations secretary-general António Guterres warned[24] of a "horrifying" surge in global domestic abuse linked to pandemic lockdowns, as calls to helplines in some countries had by then already doubled.[25]

Child abuse, meanwhile, is less likely to be detected and reported thanks to virtual schooling. There have been signs of rising child abuse, though, including a British study that found a shocking 1,493 percent rise in the incidence of abusive head trauma among children during the first month of the lockdown, compared with the same time period in the previous three years.[26]

Children are also in danger of falling behind socially and developmentally, even if they're not exposed to direct abuse. According to one report, scholastic achievement gaps widened in the US and early literacy among kindergarteners saw a sharp decline in 2020.[27]

According to *The Economist*, American children over the age of 10 cut physical activity by half during the lockdown, spending most of their time playing video games and eating junk food.[28] Indeed, closing parks and beaches right along with small businesses and schools were undoubtedly among the most ignorant and destructive pandemic measures of all.

Preventing healthy people from working and upending everyone's lives has also (as expected) resulted in a massive rise in suicide—including among children—and abnormal spikes became apparent within weeks of the initial lockdowns. In September 2020, Cook Children's Medical Center in Fort Worth, Texas, admitted a record number of 37 pediatric patients who had tried to commit suicide.[29]

In Japan—which didn't even implement lockdowns—government statistics reveal that more people died from suicide in the month of October than have died from COVID-19 all year.[30] While only 2,087 Japanese had died from COVID-19 as of November 27, 2020, the suicide toll in October alone was 2,153. Women make up the lion's share of suicides, and hotlines are also reporting that women are confessing thoughts of killing their children out of sheer desperation.

It should be obvious to anyone paying attention that the pandemic is being prolonged and exaggerated for a reason, and it's not because there's concern for life. Quite the contrary. It's a ploy to quite literally enslave the global population within a digital surveillance system—a system so unnatural and inhumane that no rational population would ever voluntarily go down that road.

How They Engineered Panic

Establishment health officials, virologists, and genetic engineers are funded by military biodefense/biowarfare programs, Big Pharma, and government. They contend the SARS-CoV-2 virus is so infectious and dangerous that there are currently no existing medical drugs, treatment protocols, supplements, natural herbs, health practices, or dietary or lifestyle changes that can strengthen your natural immune system and protect you from serious illness, hospitalization, or even death from the virus.

The authorities tell you there is no choice but to follow orders, obey the rules of mask wearing and lockdowns, and wait for Big Pharma to deliver at "Warp Speed" their inadequately tested, genetically engineered vaccines. This orchestrated panic narrative is a Big Lie, meant to keep us, the global underclass, in line, locked down, and obeying authority.

With the body politic divided, misinformed, censored, and living in panic, the globalists, the world economic elite, can consolidate their wealth and power beyond anything the world has ever seen, hiding behind the excuse that they are safeguarding public health, mitigating climate change, and eliminating poverty and unemployment. In the shadow of the Big Lie, our only hope is to spread the truth, resist, get organized, and stop this tyrannical New World Order.

Know That You Are Not Powerless

It is essential for your survival to reject the panic narrative, move beyond fear, and take charge of your mental and physical health. We must expose the manipulated calibrations and built-in shortcomings of the PCR lab tests that are creating an artificial sense of panic.

It is imperative to understand the statistics on death and hospitalizations in a manner that creates knowledge, not irrational fear. Youth and those who are metabolically healthy are typically not at risk. And fortunately, there are a large number of tried and proven means to protect the most vulnerable.

We can prevent the spread of COVID-19 and mitigate the effects of the virus by improving public health, which includes simple strategies such as eliminating processed food in our diets, making sure that healthy organic foods are available to everyone, and promoting exercise. The solution is to move beyond fear and isolation and educate yourself, as well as those you love and care for, to understand that you are not powerless.

As natural health advocate Nate Doromal reminds us: "Covid-19 is not going away. Despite prolonged lockdowns and widespread mask mandates, Covid-19 is still present in our society and cases continue throughout the country. Even the much-discussed Covid-19 vaccine is not a panacea; authorities say it will not prevent transmission and there are outstanding safety concerns amongst the leading Covid-19 vaccine candidates. The key lies in making ourselves stronger."[31]

Truth be told, we can make our bodies stronger, can make our immune systems more powerful, and can even reverse chronic preexisting conditions. It's never too late to take the steps to improve your health and make yourself more resilient to infectious diseases like COVID-19.[32]

While proponents of the official story continue to denigrate and slander COVID-19 critics, including the authors of this book, as "anti-science, anti-vaccine conspiracy theorists," the evidence points to SARS-CoV-2 being a weaponized, lab-engineered, highly transmissible biological trigger that magnifies and exacerbates preexisting chronic diseases and comorbidities. COVID-19 presents basically no threat to children, youth, and students, and very little threat to people in good health of any age, unlike the Spanish flu of 1918.

People over 65 years old who are metabolically unhealthy and/or have low vitamin D levels, as well as those with serious preexisting chronic disease such as obesity, diabetes, heart disease, cancer, lung disease, kidney disease, dementia, and hypertension, need to safeguard their health and strengthen their bodies' ability to fight off disease by taking precautions that minimize exposure to the SARS-CoV-2 virus, as well as other viruses such as the seasonal flu.

For those in nursing homes or hospitals, special precautions are also necessary. As the Great Barrington Declaration, signed by tens of thousands of doctors and scientists around the world, points out:

> *Adopting measures to protect the vulnerable should be the central aim of public health responses to COVID-19. By way of example, nursing homes should use staff with acquired immunity and perform frequent PCR testing of other staff and all visitors. Staff rotation should be minimized.*
>
> *Retired people living at home should have groceries and other essentials delivered to their home. When possible, they should meet family members outside rather than inside. A comprehensive and detailed list of measures, including approaches to multi-generational households, can be implemented, and is well within the scope and capability of public health professionals.*
>
> *Those who are not vulnerable should immediately be allowed to resume life as normal. Simple hygiene measures, such as hand washing and staying home when sick should be practiced by everyone to reduce the herd immunity threshold. Schools and universities should be open for in-person teaching.*
>
> *Extracurricular activities, such as sports, should be resumed. Young low-risk adults should work normally, rather than from home. Restaurants and other businesses should open. Arts, music, sport and other*

cultural activities should resume. People who are more at risk may partic-ipate if they wish, while society as a whole enjoys the protection conferred upon the vulnerable by those who have built up herd immunity.[33]

The continuation of school closures, lockdowns, and other extreme mea-sures that fall hardest on low-income groups, minority communities, small businesses, and children are counterproductive and wrong. We need to reduce public panic and political polarization and have a serious, society-wide discus-sion on the origins, nature, virulence, prevention, and treatment of COVID-19.

The Panic Narrative Is Built upon Faulty Tests, Misleading Statistics, and Shoddy Science

There are several major aspects of the official "scientific" narrative on the nature, infectivity, and virulence of COVID-19 that are deliberately mislead-ing and spreading panic among the public. These include the use of faulty, miscalibrated PCR lab tests that artificially inflate the number of COVID-19 cases, which we reviewed in chapter 4.

The fact is, a vast majority of those who test positive for SARS-CoV-2 remain asymptomatic and are highly unlikely to spread the disease to others. They simply aren't sick. The PCR test is merely picking up inactive (non-infectious) viral particles.

In one study, which looked at pregnant women admitted for delivery, 87.9 percent of the women who tested positive for the presence of SARS-CoV-2 had no symptoms.[34] Another study looked at a large homeless shelter in Bos-ton. Of 408 people tested, 147 (36 percent) were positive, yet symptoms were conspicuously absent. Cough occurred in only 7.5 percent of cases, shortness of breath in 1.4 percent, and fever in 0.7 percent. All symptoms were "uncom-mon among COVID-positive individuals," the researchers noted.[35]

A study in *Nature Communications* assessed the risk posed by asymptom-atic people by looking at the data from a mass screening program in Wuhan, China. The city had been under strict lockdown between January 23 and April 8, 2020. Between May 14 and June 1, 2020, 9,899,828 residents of Wuhan city over the age of six underwent PCR testing. Of these, 9,865,404 had no previous diagnosis of COVID-19 and 34,424 were recovered COVID-19 patients. In all, there were zero symptomatic cases and only 300 asymptomatic cases detected. (The overall detection rate was 0.3 per 10,000.) Importantly, not a single one of the 1,174 people who had been in close contact with an asymptomatic individual tested positive.

Additionally, of the 34,424 participants with a history of COVID-19, 107 individuals (0.310 percent) tested positive again, but none were symptomatic. As noted by the authors, "Virus cultures were negative for all asymptomatic positive and repositive cases, indicating no 'viable virus' in positive cases detected in this study." Interestingly, when they further tested asymptomatic patients for antibodies, they discovered that 190 of the 300 (63.3 percent) had actually had a "hot" or productive infection resulting in the production of antibodies, yet none of their contacts had been infected. In other words, even though asymptomatics were (or had been) carriers of apparently live virus, they *still* did not transmit it to others.[36]

If positive test results tell us nothing about the actual prevalence of disease and its spread, why are we mass-testing? Of course, if PCR testing is unreliable, then statistics and public statements by vaccine manufacturers on the efficacy of their vaccines to prevent or cure COVID-19 are also invalid, since they "prove" efficacy using these tests.

Another misleading practice is to conflate statistics on deaths. As reviewed in the previous chapter, 94 percent of so-called COVID-19 deaths were people who died *with* COVID-19, as they had other preexisting chronic diseases or comorbidities.[37] The idea that COVID-19 is a lethal pandemic is also disproven by all-cause mortality statistics, which show mortality has remained steady during 2020 and doesn't veer from the norm.[38]

Other fearmongering tactics include public statements exaggerating the threat of COVID-19 to children, youth, and students as well as the risk of youth to spread COVID-19 to teachers and older adults in general. Even Anthony Fauci now admits that students pose little or no threat to teachers or older adults and that schools should be reopened.[39]

Equating Faulty PCR Positive Test Results as "Cases" of Active Infections of COVID-19

While the death toll was initially the driving fear tactic, it quickly shifted to the dubious claim that there are "increasing cases" of COVID-19, including among the young. These news reports or public health proclamations are often accompanied by ominous graphs, always trending upward, with dire warnings of a "second or third wave" of mass hospitalizations and deaths being imminent if people don't hunker down, obey authority, and isolate themselves as thoroughly as they did in the early stages of the pandemic.

Hardly ever do these stories mention that 10 times as many people are now being tested as were being tested during the early stages of the pandemic, or that there is mounting evidence of false positives, caused by laboratory over-magnification of what are supposedly viral samples from nasal or throat swabs.

In the fine print of these alarming news reports, there are often admissions that while actual deaths from COVID-19 have declined, we can expect massive deaths if people stop wearing masks or resume a semblance of normal life. The upbeat news of these scare articles is that the danger of infection and death will eventually subside once everyone gets vaccinated.

But we need to ask ourselves: What do these experts and media outlets actually mean by an increasing number of "cases" of COVID-19?

Do they mean that more people than ever are getting seriously ill and are dying from COVID-19? If so, why do official statistics from the CDC and other public health databases show declining numbers of deaths from COVID-19 across the US and the world, even when flu and pneumonia cases are misleadingly counted as COVID-19 cases?[40]

Or does it simply mean that more and more people, especially now young people, are being tested and end up with a positive test result? And if so, what does *that* actually mean? As former *New York Times* reporter Alex Berenson points out in his book *Unreported Truths About COVID-19 and Lockdowns*: "A 'case' of coronavirus points only to a positive test result . . . It does not mean that a person will become sick—much less that he or she will be hospitalized, need intensive care, or die."[41]

At present, the polymerase chain reaction test is the primary method used to test people for COVID-19. The problem with that is twofold. First of all, the PCR test cannot distinguish between inactive viruses and "live" or reproductive ones.[42] This is a crucial point, since inactive and reproductive viruses are not interchangeable in terms of infectivity. If you have a nonreproductive virus in your body, you will not get sick and you cannot spread it to others. For this reason, the PCR test is grossly unreliable as a diagnostic tool.

Second, many if not most laboratories amplify the RNA collected far too many times, which results in healthy people testing positive. In order for the PCR test to be of any use whatsoever, in terms of diagnosing COVID-19, labs would need to considerably reduce the number of amplification cycles used.

Here's what you need to understand about the PCR test: The PCR swab collects RNA from your nasal cavity. This RNA is then reverse-transcribed into DNA. Due to its tiny size, it must be amplified to become discernible. Each round of amplification is called a cycle, and the number of amplification cycles used by any given test or lab is called a cycle threshold (CT). The higher the CT, the greater the risk that insignificant sequences of viral DNA end up being magnified to the point that the test reads positive even if your viral load is extremely low or the virus is inactive and poses no threat to you or anyone else.

Many scientists have noted that anything over 35 cycles is scientifically indefensible.[43] Even Dr. Anthony Fauci, a leading proponent of gain-of-function experiments and mandatory vaccines, has admitted that the chances of a positive PCR result being accurate at 35 cycles or more "are minuscule."[44]

A September 28, 2020, study[45] in *Clinical Infectious Diseases* revealed that when you run a PCR test at a CT of 35 or higher, the accuracy drops to 3 percent, resulting in a 97 percent false positive rate. Yet tests recommended by the World Health Organization are set to 45 cycles,[46] and the US Food and Drug Administration and the US Centers for Disease Control and Prevention recommend running PCR tests at a CT of 40.[47] The question is why, considering the consensus is that CTs over 35 render the test useless. When labs use these excessive cycle thresholds, you clearly end up with a grossly overestimated number of positive tests, so what we're really dealing with is a "casedemic"—an epidemic of false positives.[48]

As noted by author and investigative journalist Jon Rappoport:

> *All labs in the US that follow the FDA guideline are knowingly or unknowingly participating in fraud. Fraud on a monstrous level, because . . . Millions of Americans are being told they are infected with the virus on the basis of a false positive result, and . . . The total number of COVID cases in America—which is based on the test—is a gross falsity. The lockdowns and other restraining measures are based on these fraudulent case numbers . . .*[49]

Now, if CTs above 35 are scientifically unjustified, just how low of a CT should be used? Quite a few studies have investigated this, so there's no shortage of data at this point. The fact that the WHO, FDA, and CDC still have not changed their CTs downward in light of all these data tells us they're not interested in getting an accurate picture of the infection rate.

For example, an April 2020 study in the *European Journal of Clinical Microbiology and Infectious Diseases* showed that to get 100 percent confirmed real positives, the PCR test must be run at 17 cycles. Above 17 cycles, accuracy drops dramatically.[50]

By the time you get to 33 cycles, the accuracy rate is a mere 20 percent, meaning 80 percent are false positives. Beyond 34 cycles, your chance of a positive PCR test being a true positive shrinks to zero. According to a December

3, 2020, systematic review published in the journal *Clinical Infectious Diseases*, no live viruses could be found in cases where a positive PCR test had used a CT above 24.[51]

What these studies show, then, is that if you actually have symptoms of COVID-19 and test positive using a PCR test that was run at 35 amplification cycles or higher, then you're probably infected and likely infectious. However, if you do *not* have symptoms, yet test positive using a PCR test run at 35 CTs or higher, then it is likely a false positive and you pose no risk to others as you're unlikely to carry any live virus. In fact, provided you're asymptomatic, you're unlikely to be infectious even if you test positive with a test run at 24 CTs or higher. This supports the findings presented earlier in this chapter, which show that asymptomatic people (those who test positive but have no symptoms) are extremely unlikely to transmit live virus to others.

According to Stephen A. Bustin, professor of molecular medicine and a world-renowned expert on the PCR test, when you get a positive result using a CT of 35 or higher, you're looking at the equivalent of a single copy of viral DNA, and the likelihood of that causing a health problem is minuscule.[52]

If you want to frighten people, sell more PCR tests, or reinitiate lockdowns, all you have to do is require more testing and calibrate the tests so that people who are not sick or contagious *appear* to be infected and able to spread COVID-19. Considering how few governments have taken action to remedy this artificial inflation of COVID-19 cases—which is easy enough—we really have to wonder whether it's part of a global agenda to keep the fear level elevated.

In December 2020, Florida became the first US state to require labs to report the cycle threshold used for their PCR tests.[53] In Europe, meanwhile, a court in Portugal has ruled that the PCR test is "not a reliable test for SARS-CoV-2, and therefore any enforced quarantine based on the results is unlawful."[54] China addressed the PCR problem by simply stopping testing people for COVID-19 except for those actually exhibiting symptoms.

As for how to properly confirm a COVID-19 diagnosis, a review of COVID-19 PCR testing emphatically states:

> To determine whether the amplified products are indeed SARS-CoV-2 genes, biomolecular validation of amplified PCR products is essential. For a diagnostic test, this validation is an absolute must.
> Validation of PCR products should be performed by either running the PCR product in a 1% agarose-EtBr gel together with a size indicator (DNA ruler or DNA ladder) so that the size of the product can be

estimated. The size must correspond to the calculated size of the ampli-fication product. But it is even better to sequence the amplification product. The latter will give 100% certainty about the identity of the amplification product. Without molecular validation one cannot be sure about the identity of the amplified PCR products . . . *[emphasis ours].*[55]

A similar argument has been made for confirmatory molecular sequencing in a petition by the European Medicines Agency to put a halt to COVID-19 vaccine trials that are using misleading PCR tests.[56]

Fatal Errors Found in Paper on Which PCR Testing Is Based

On November 30, 2020, a team of 22 international scientists published a scathing review[57] challenging the scientific paper on PCR testing for SARS-CoV-2 written by (among others) Christian Drosten, PhD, and Victor Corman.[58] The Corman-Drosten paper had been quickly accepted by the World Health Organization, and the workflow described therein was adopted as the standard across the world.

According to Reiner Fuellmich, founding member of the German Corona Extra-Parliamentary Inquiry Committee (Außerparlamentarischer Corona Untersuchungsausschuss),[59] Drosten is a key culprit in the COVID-19 pandemic hoax.

The scientists demanded that the Corman-Drosten paper be retracted due to several "fatal errors," one of which is the fact that it was written (and the test itself developed) before any viral isolate was available. All they used was the genetic sequence published online by Chinese scientists in January 2020.

The fact that the paper was published a mere 24 hours after it was submitted also suggests it didn't even undergo peer review. In an *UncoverDC* interview, Kevin Corbett, PhD, one of the 22 scientists who demanded the paper's retraction, stated:

> *Every scientific rationale for the development of that test has been totally destroyed by this paper. It's like Hiroshima/Nagasaki to the COVID test.*
>
> *When Drosten developed the test, China hadn't given them a viral isolate. They developed the test from a sequence in a gene bank. Do you see? China gave them a genetic sequence with no corresponding viral isolate. They had a code, but no body for the code. No viral morphology.*

In the fish market, it's like giving you a few bones and saying "that's your fish." It could be any fish . . . Listen, the Corman–Drosten paper, there's nothing from a patient in it. It's all from gene banks. And the bits of the virus sequence that weren't there they made up. They synthetically created them to fill in the blanks . . .

There are 10 fatal errors in this Drosten test paper . . . But here is the bottom line: There was no viral isolate to validate what they were doing . . . There have since been papers saying they've produced viral isolates. But there are no controls for them. The CDC produced a paper in July . . . where they said: "Here's the viral isolate." Do you know what they did? They swabbed one person. One person, who'd been to China and had cold symptoms. One person. And they assumed he had [COVID-19] to begin with. So, it's all full of holes, the whole thing.[60]

The conclusion of the review reads, in part:

A decision to recognize the errors apparent in the Corman–Drosten paper has the benefit to greatly minimize human cost and suffering going forward. Is it not in the best interest of Eurosurveillance to retract this paper? Our conclusion is clear. In the face of all the tremendous PCR-protocol design flaws and errors described here, we conclude: There is not much of a choice left in the framework of scientific integrity and responsibility.[61]

The critique against PCR testing is further strengthened by the November 20, 2020, study in *Nature Communications*, discussed earlier in this chapter, which found no viable virus in PCR-positive cases at all.[62]

Class-Action Lawsuits Against Fraudulent SARS-CoV-2 Testing

In early October 2020 an international team of class-action lawyers, led by Reiner Fuellmich, announced they will soon be filing massive lawsuits against a number of governments for utilizing imprecise PCR and antibody tests—which generate huge profits for Big Pharma as well as vaccine and testing companies—and then knowingly using the data from these faulty tests to justify lockdowns and suspensions of basic civil liberties, resulting in massive damage to public health, businesses, and citizens.[63]

As Fuellmich states, PCR tests, according to the leaflets that accompany the test kits, should not be considered true diagnostic tests for the presence of disease. Even the CDC admitted in a July 13, 2020, statement that PCR tests "May not necessarily indicate the presence of an infectious virus," "May not prove that a SARS-CoV-2 fragment is the cause of clinical symptoms," and cannot rule out diseases caused by other bacterial or viral pathogens.[64]

A September 20, 2020, "Open Letter from Medical Doctors and Health Professionals to All Belgian Authorities and All Belgian Media" reiterates some of the serious shortcomings of the PCR tests that are currently relied on to make the alarming claim that cases are rising across the US, Europe, and the world:

> *The use of the non-specific PCR test, which produces many false positives, showed an exponential picture. This test was rushed through with an emergency procedure and was never seriously self-tested. The creator expressly warned that this test was intended for research and not for diagnostics.*
>
> *The PCR test works with cycles of amplification of genetic material—a piece of genome is amplified each time. Any contamination (e.g. other viruses, debris from old virus genomes) can possibly result in false positives.*
>
> *The test does not measure how many viruses are present in the sample. A real viral infection means a massive presence of viruses, the so-called virus load.* If someone tests positive, this does not mean that that person is clinically infected, is ill or is going to become ill *[emphasis ours].*[65]

Since a positive PCR test cannot reliably or automatically indicate active infection or infectivity, there's absolutely no justification for the social measures taken, as they are based solely on these tests.

On January 20, 2021, roughly an hour after Joe Biden's inauguration as the 46th president of the United States, the World Health Organization suddenly and out of the blue lowered the recommended PCR cycle threshold (CT),[66] which automatically guarantees that the number of "cases," i.e., positive PCR test results, will plummet. The next day, January 21, 2021, President Biden announced he would be reinstating the US' financial support for the WHO.[67] Dr. Meryl Nass explains, "The WHO instructed PCR test users and manufacturers on December 14[68] and again on January 20[69] that PCR cycle thresholds needed to come down. The December 14 guidance stated WHO's concern regarding 'an elevated risk for false SARS-CoV-2 results' and pointed to 'background noise which may lead to a specimen with a high-cycle threshold value result being *[incorrectly]* interpreted as a positive result.'"[70] As PCR cycles have been reduced, new "cases" dropped 60 percent from 250,000 new cases per day to 100,000, in January, while hospitalization rates[71] associated with COVID dropped from a high of 132,500 Americans on January 6 to 71,500 on February 12.[72] Of course, health authorities and the mass media have attributed this sharp drop in US "cases" and hospitalizations to vaccines, masks, and social distancing, rather than WHO-mandated recalibration of PCR tests.

COVID-19 Rules Mark "Hysterical Slide into Police State"

The dangers of fearmongering are summed up well by British Supreme Court judge Lord Sumption in a March 30, 2020, interview with *The Post*. Sumption warned that COVID-19 rules are paving the way for despotism—the exercise of absolute power in a cruel and oppressive manner.

> *The real problem is that when human societies lose their freedom, it's not usually because tyrants have taken it away. It's usually because people willingly surrender their freedom in return for protection against some external threat. And the threat is usually a real threat but usually exaggerated.*
>
> *That's what I fear we are seeing now. The pressure on politicians has come from the public. They want action. They don't pause to ask whether the action will work. They don't ask themselves whether the cost will be worth paying. They want action anyway. And anyone who has studied history will recognize here the classic symptoms of collective hysteria.*
>
> *Hysteria is infectious. We are working ourselves up into a lather in which we exaggerate the threat and stop asking ourselves whether the cure may be worse than the disease.*[73]

Indeed, in just a few short months, we dramatically shifted from a state of freedom to a state of totalitarianism, and the way that was done was through social engineering, which of course involves psychological manipulation.

Censoring and propaganda are but two strategies that shape and mold a population. Psychiatry professor Albert Biderman's "chart of coercion"[74] also includes the following methods, all of which can be clearly related to the COVID-19 response:

Isolation techniques—Quarantines, social distancing, isolation from loved ones, and solitary confinement.

Monopolization of perception—Monopolizing the 24/7 news cycle, censoring dissenting views, and creating barren environments by closing bars, gyms, and restaurants.

Degradation techniques—Berating and shaming (or even physically attacking) those who refuse to wear masks or social distance, or generally choose freedom over fear.

Induced debility—Being forced to stay at home and not be able to exercise or socialize.

Threats—Threatening with the removal of your children, prolonged quarantine, closing of your business, fines for noncompliance with mask and social distancing rules, forced vaccination, and so on.

Demonstrating omnipotence/omniscience—Shutting down the whole world, claiming scientific and medical authority.

Enforcing trivial demands—Examples include family members being forced to stand six feet apart at the bank even though they arrived together in the same car, having to wear a mask when you walk into a restaurant even though you can remove it as soon as you sit down, or having to wear a mask when walking alone on the beach.

Occasional indulgence—Reopening some stores and restaurants but only at a certain capacity, for example. Part of the coercion plan is that indulgences are given, then taken away again.

It is time to ask ourselves some very pressing questions. Is it reasonable to expect government to eliminate *all* infection and *all* deaths? They've proven they cannot, yet we keep relinquishing more and more freedoms and liberties because they claim doing so will keep everyone safer. It's an enticing lie, but a lie nonetheless.

Sooner or later everyone must decide which is more important: human rights and constitutional freedoms, or false security. The good news is that many are starting to see the writing on the wall; they're starting to see we've been had, and are starting to choose liberty over brutal totalitarianism in the name of public health.

Remember what Ben Franklin once said: "Those who would give up essential Liberty, to purchase a little temporary safety, deserve neither liberty nor safety."

The truth is, the technocrats have no intention of ever letting us go back to normal. The plan is to alter society *permanently*. Part of that alteration is the removal of civil liberties and human rights, which is now happening at breakneck speed.

CHAPTER SIX

Protecting Yourself from COVID-19

By Dr. Joseph Mercola

It's undeniable that for the past 100 years we have had an epidemic of chronic disease that is largely related to a radical increase in the consumption of processed foods. General health and mortality statistics in general make this easy to see, but so do COVID-19 statistics specifically. As illustrated in chapter 4, the overwhelming majority of people with severe COVID outcomes have not just one, but several underlying health conditions or comorbidities.

We talked about many of the big ones, including insulin resistance, obesity, diabetes, and hypertension in that chapter, but lung disease, cancer, and dementia are also to blame. Why are we all so sick in the first place?

In many ways, we have Big Ag, Big Food, and Big Pharma to thank for the COVID-19 pandemic, seeing how these industries are responsible for the epidemic of chronic ill health that the SARS-CoV-2 infection piggybacks on. While these industries have set us up to believe it's normal to be chronically ill, there's nothing normal, or even acceptable, about it. There are simple, safe, highly effective, and relatively inexpensive strategies that can boost your immune system and defend yourself not only from COVID-19 but from virtually all chronic disease as well.

Blame Big Food, Big Ag, and Big Pharma

In chapter 4 we reviewed how the majority of COVID-19 deaths aren't caused by the virus itself. Rather, they're caused by comorbidities that are a direct result of a highly processed diet—put in place, promoted, and maintained by Big Food and Big Ag—and an over-reliance on Big Pharma's solutions that focus only on treating symptoms.

The good news is you can reclaim ownership of your health from these corporate threats. In this chapter we'll review the optimal diet for health and well-being, supplements to combat chronic disease and viral infection, and additional strategies to keep you resistant to illness and infection of all kinds.

But first I want to just give an overview of how we got here. I'm going to call out Big Food and Big Ag separately below (we'll also cover Big Pharma more exclusively in chapter 7). As you read, it will become even clearer just how these corporate interests have worked to create both the chronic disease epidemic and the COVID-19 pandemic.

If there is a silver lining to this pandemic, it's that it's pulling back the curtain, showing the world that Big Ag, Big Food, and Big Pharma don't have your best interest at heart. It's showcasing that—although they don't want you to believe this—healthy lifestyles provide essential immunity to these types of infections and disastrous health consequences.

You *can* take control of your health and strengthen your own innate immune system; you don't have to rely on medications or vaccines. Because here's the deal: Natural immunity is lifelong; artificial immunity from synthetic and potentially harmful vaccines is not. Ultimately, the way we eradicate COVID-19 is by improving the general health of the public, and to do that, we need to stress the importance of a health-promoting diet.

How the Food Industry Deceives You into Eating Unhealthy Foods

Underlying health conditions like obesity, heart disease, and diabetes are the real pandemic here. Obesity alone doubles your risk of being hospitalized for COVID-19 and raises your risk of death anywhere from 3.68 times to 12 times, depending on your level of it. Processed food (loaded with industrially processed vegetable oils) and soft drinks (chock-full of sugar) are key culprits in the development of chronic disease, and therefore have a key role to play in COVID-19 hospitalizations and deaths as well.

Beyond the addictive potential of these foods and beverages is the marketing used to sell them, which further entices Americans to buy and consume more. This includes things like positioning junk foods at eye level on grocery store displays.

In an editorial published in the *BMJ*, three researchers cited the role of the food industry in driving up rates of obesity and ultimately causing more COVID-19 deaths.[1] According to the authors, "It is now clear that the food industry shares the blame not only for the obesity pandemic but also for the severity of COVID-19 disease and its devastating consequences."

To address this underlying connection, they called on the food industry to immediately cease promoting unhealthy food and drinks, and for governments to force reformulation of junk foods to better support health.

Yet even in the midst of the COVID-19 pandemic, multinational food and beverage corporations are interfering with public policy and influencing the development of dietary guidelines. According to a report published by the campaign group Corporate Accountability,[2] more than half of those appointed to the 2020 Dietary Guidelines Advisory Committee (DGAC) have ties to the International Life Sciences Institute (ILSI), a not-for-profit organization established by a Coca-Cola executive 40 years ago[3] and funded by multinational junk food companies such as Coca-Cola, PepsiCo, McDonald's, General Mills, and Cargill. To protect public health, this conflicted influence must be curbed, the report warned.

DGAC is supposed to be an independent committee that reviews scientific evidence and provides a report to help develop the dietary guidelines for Americans. As the go-to source for nutrition advice in the US, DGAC dictates what more than 30 million US schoolchildren eat at school and drives the nutritional advice given to new mothers, seniors, veterans, and other beneficiaries of nutritional education and meals offered by the federal government.

Its extensive ties to ILSI all but ensure that the DGAC is anything but independent. ILSI has been exposed as a shill for the junk food industry, and internal documents have revealed ILSI embedded itself in public health panels across Europe and the United Nations in an effort to promote its own industry-focused agenda and raise profits at the expense of public health worldwide.[4]

The Corporate Accountability report further examined ILSI's "revolving doors and conflicts of interest" with other critical government policy processes, including updating national food composition databases.

Yet despite being called out publicly during an unprecedented challenge to public health, junk food giants' influence continues to be felt around the globe. "Even in times of crisis, such as today's COVID-19 pandemic, ILSI's backers feel no scruples lobbying for the bottom line," Corporate Accountability stated, adding:

> In India, despite potential consequences to the health and well-being of workers and the community, corporations including Coca–Cola, PepsiCo and Nestlé, have submitted letters to the government requesting food and beverage manufacturing be exempt from the lockdown, and be considered an "essential service." Not providing immune–suppressing sugar-sweetened beverages during this time may . . . prove the more essential service these corporations can provide in this time and beyond.[5]

How Junk Food Is Causing Increased COVID-19 Deaths

Processed foods—made from components extracted from other foods, such as soy protein isolate, or factory-farmed meat, then loaded with salt, sugar, and/or industrially processed vegetable oil (likely all three)—are designed to be appealing, hyperpalatable, and habit-forming, thanks to additives, crafty packaging and marketing, and a high "convenience" factor.

Yet processed foods give you calories without the vitamins, minerals, live enzymes, micronutrients, healthy fats, and high-quality protein your body needs. Processed foods increase how fast you eat and delay how "full" you feel, leading to obesity and metabolic dysfunction.

They also increase the risk of conditions such as obesity, cancer, type 2 diabetes, and cardiovascular disease, which heighten your risk of COVID-19, and compromise your gut microbiome, which plays a crucial role in your body's immune response to infection and in maintaining overall health.

Even before SARS-CoV-2 surfaced, processed foods were a really bad idea. In fact, eating over four servings of processed foods daily was found to increase your risk of premature death by 62 percent in a 2019 study.[6] During the current pandemic, their toxic influence is exponentially magnified. And since diet-related comorbidities are responsible for 94 percent of all COVID-19-related deaths,[7] taking control of your diet is a really simple, commonsense strategy to lower the risks associated with this infection.

London-based cardiologist Dr. Aseem Malhotra has been among those warning that poor diet can increase your risk of dying from COVID-19. He told BBC that processed foods make up more than half the calories consumed by the British. He tweeted, "The government and public health England are ignorant and grossly negligent for not telling the public they need to change their diet now."[8]

On the brighter side, he also states that eating nutritious foods for even one month could help you lose weight, put type 2 diabetes into remission, and improve your health considerably, so you'll have a much better chance of survival should you contract COVID-19.[9] Malhotra also told the food industry to "stop mass-marketing and selling processed food."

Dr. Robert Lustig, emeritus professor of pediatrics in the division of endocrinology at the University of California–San Francisco, has also been outspoken about the connection between diet and COVID-19 risks, stating:

> COVID . . . doesn't distinguish who it infects. But it does distinguish who it kills. Other than the elderly, it's those who are Black, obese, and/or have pre-existing conditions. What distinguished these three

demographics? Ultra-processed food. Because ultra-processed food sets
you up for inflammation, which COVID-19 is happy to exploit. . . .
Time to rethink your menu.[10]

Processed Foods Are Especially Harmful to Poor Communities

People living in poverty, whether in developing or advanced countries, are especially vulnerable to health problems from processed foods and COVID-19. According to Malhotra: "[T]he disproportionate numbers of those from black and ethnic minority backgrounds succumbing to the virus may in part be explained by a significantly increased risk of chronic metabolic disease in these groups."[11]

Even before the COVID-19 pandemic, food giants have targeted those with low incomes with aggressive marketing of ultra-processed foods. Following initiatives by Brazil to fight the trend, Ecuador, Uruguay, and Peru have urged citizens to avoid processed foods in favor of natural foods.[12]

Food deserts further the dietary exploitation of the poor. The USDA defines a *food desert* as a low-income tract where many residents do not have easy access to a supermarket or large grocery store.[13] In addition to a lack of food outlets offering healthy food, residents' lack of transportation to get to stores is a big factor. Residents who have to walk with their groceries or take the bus can carry fewer groceries, and transporting perishable items can be a major obstacle to obtaining certain health foods.

Your Diet Can Radically Improve Your Immune Function

You can combat insulin resistance and obesity, as well as prevent most chronic diseases, with a healthy diet. Of course, what you don't eat is just as important. This is why eliminating as many processed foods and fast foods as possible is your first priority.

But even if you are only eating healthy whole foods, it is important to understand that 9 out of 10 people are metabolically unhealthy. How do you know if you are in that majority? If you answer yes to any of the four questions below, then there is a good chance you are, and the more yes answers, the higher the likelihood that you are metabolically unhealthy.

- Do you have diabetes?
- Do you have high blood pressure?
- Are you overweight?
- Are your fasting triglycerides higher than your HDL?

If you are metabolically unhealthy, then it would be wise to limit your net carbs (total carbohydrates minus fiber) to approximately 50 grams per day or about 15 percent of your total calories. The best way to calculate this would be to use a free desktop app called Cronometer, which is further explained in more detail in the next section on fat.

Once you recover your metabolic flexibility, resolve your insulin resistance, and are at or near your ideal weight, you can start cycling carbs back in. Depending on how much exercise you engage in, you could easily triple your carb intake. It is best to do this cyclically. For some, that could be as frequent as every other day. For others, it might be once or twice a week.

For a much more in-depth guide to cyclical ketosis and guidance on how to prevent and even reverse the chronic diseases that plague our society and make so many of us vulnerable to COVID-19, refer to my previous two books, *Fat for Fuel* and *KetoFast*.

The Most Dangerous Fat of All

My next book will focus on omega-6 linoleic acid (LA), which makes up the bulk—about 90 percent—of the omega-6 consumed and is the primary contributor to nearly all chronic diseases. While excess sugar is certainly bad for your health and should typically be limited as discussed in the section above, it doesn't cause a fraction of the oxidative damage that LA does. While an essential fat, when consumed in excessive amounts, LA actually acts as a metabolic poison that impairs the function of your mitochondria and triggers the destruction of cells.

Its adverse effects are primarily due to the fact that it's a highly perishable fat, prone to oxidation. As the fat oxidizes, it breaks down into by-products such as advanced lipid oxidation end products (ALEs) and oxidized LA metabolites (OXLAMs), which are extraordinarily harmful even in exceedingly small quantities. One type of advanced lipid oxidation end product is 4HNE, a mutagen known to cause DNA damage. Studies have shown there's a definite correlation between elevated levels of 4HNE and heart failure, for example. The amount of LA in adipose tissue and platelets is positively associated with coronary artery disease. LA breaks down into 4HNE faster when the oil is heated, which is why cardiologists recommend avoiding fried foods. LA intake and the subsequent ALEs and OXLAMS produced also play a significant role in cancer.

Processed vegetable oils are a primary source of LA, but even foods conventionally hailed for their health benefits, such as olive oil, chicken, and farmed salmon, contain it. For clarity, while you do need some LA, it becomes a major problem if consumed in excess, and the problem is that virtually everyone is eating excessive LA and completely unaware of its deleterious health consequences. Importantly, simply increasing your omega-3 intake is not the answer here, as it will not counteract the damage done by excessive LA. To prevent problems, you really need to minimize your intake of omega-6 fats.

Linoleic Acid Intake May Affect COVID-19 Outcomes

Your LA intake may even have a direct impact on your COVID-19 risk. According to a September 2020 report in the journal *Gastroenterology*, your risk of dying from COVID-19 actually appears to be heavily influenced by the amount of unsaturated fats you eat, as they play a role in organ failure.[14] In summary, higher intakes of polyunsaturated fats (PUFAs), primarily LA, resulted in a greater risk of severe COVID-19, while higher intake of saturated fat lowered the risk.

According to the authors, unsaturated fats "cause injury [and] organ failure resembling COVID-19." More specifically, unsaturated fats are known to trigger lipotoxic acute pancreatitis, and the sepsis and multisystem organ failure seen in severe cases of COVID-19 greatly resembles this condition.

They noted that hypocalcemia (lower-than-average levels of calcium in your blood or plasma) and hypoalbuminemia (low albumin in your blood) are observable early on in patients with severe COVID-19. Low arterial partial pressure of oxygen and percentage of oxygen ratios were also associated with higher levels of unbound fatty acid levels in patients' blood. Unsaturated fats may also cause vascular leakage, inflammatory injury, and arrhythmia during severe COVID-19.

In tests on mice, animals given LA developed a range of conditions resembling lethal COVID-19, including hypoalbuminemia, leukopenia (low white blood cell count), lymphopenia (low lymphocyte count), lymphocytic injury, thrombocytopenia (low platelet count), hypercytokinemia (cytokine storm), shock, and kidney failure. The solution they propose is early supplementation with egg albumin and calcium, as both of these are known to bind unsaturated fats, thereby reducing injury to organs.

How to Calculate Your LA Intake with Cronometer

Considering the damage LA imparts, it's not surprising that it could play a significant role in the outcome of COVID-19. As mentioned, virtually all of the comorbidities associated with COVID-19 are diet-related, share many of the same risk factors, and can be triggered or worsened by high LA intake.

Fortunately, you won't have to spend hundreds of dollars to have your food analyzed for LA. All you need to do is accurately enter your food intake into Cronometer—a free online nutrition tracker—and it will provide you with your total LA intake. The key to accurate entry is to carefully weigh your food with a digital kitchen scale so you can enter that weight to the nearest gram.

Cronometer is free to use when you use the desktop version (www.cronometer .com.) If you feel the need to use your cellphone (which is not recommended) to enter your data, then you will need to purchase a subscription. Ideally, enter your food for the day before you actually eat it. The reason for this is simple: It's impossible to delete the food once you have already eaten it, but you can easily delete it from your menu if you find something pushes you over the ideal limit.

Once you've entered the food for the day, go to the "Lipids" section on the lower left side of the app (see figure 6.1). To determine how much LA is in your diet for that day, all you need to know is how many grams of omega-6 is present. About 90 percent of the omega-6 you eat is LA.

To find out the percentage of calories the omega-6/LA represents in your diet, go to the "Calories Summary" section (see figure 6.2). In this example, the total calorie count is 3,887. Since there are 9 calories per gram of fat, you will need to multiply the number of omega-6 grams (7.7) by 9 to obtain the total amount of omega-6 calories. In this case, that's 69.3 calories.

Next, divide the LA calories by your total calories. In this example, that would be 69.3/3887 = 0.0178. If you multiply that number by 100, or move the decimal

Lipids		
Fat	350.6 g	125%
Monounsaturated	62.8 g	No Target
Polyunsaturated	11.6 g	No Target
Omega-3	2.5 g	157%
Omega-6	7.7 g	54%
Saturated	242.7 g	n/a
Trans-Fats	6.3 g	n/a
Cholesterol	1271.8 mg	127%

Figure 6.1. Cronometer Lipids Profile.

point two spaces to the right, you will have the percentage as a whole number. In this example, it is 1.8 percent of LA. This falls within the ideal LA percentage range, which is between 1 and 2 percent of your total calorie intake. Sometime in 2021, Cronometer will update the application to automatically calculate and display the percentage of omega-6, at which point it will be even easier to use.

The Nearly Magical Eating Formula to Radically Improve Your Health

There is exciting new research that shows a special type of eating strategy—called time-restricted eating (TRE), sometimes referred to as intermittent fasting—is one of the most profoundly effective strategies to regain your metabolic flexibility.

It promotes insulin sensitivity, decreases insulin resistance, and improves blood sugar management by increasing insulin-mediated glucose uptake rates.[15] This is important not only for resolving type 2 diabetes but also high blood pressure and obesity.

It also catalyzes a very powerful cleanup tool of your body called autophagy. This is when your body removes damaged cellular parts and recycles them to make new ones. Without this process actively engaged, your body can be likened to a very old automobile that has not been maintained.

Figure 6.2. Cronometer Calories Summary.

TRE generally involves restricting your eating window to only six to eight hours, which mimics the eating habits of your ancestors. While there are a number of different TRE protocols, my preference is fasting daily for 16 to 18 hours and eating all meals within a six- to eight-hour window.

If you're new to the concept of TRE, consider starting by skipping breakfast and having your lunch and dinner within a six- to eight-hour time frame, say 11 AM to 7 PM, making sure you stop eating three hours before going to bed. It's a powerful tool that can work even in lieu of making other dietary changes. In one study, when 15 men at risk of type 2 diabetes restricted their eating to a nine-hour window, they lowered their mean fasting glucose, regardless of when the eating window commenced.[16] It is best to pick times that work for you and your family's schedule, but typically, the more time you leave between your last meal and your bedtime, the better the benefits.

Another major benefit of TRE is improved mitochondrial function. Most of your cells produce nearly all of their energy via the mitochondria. They're also responsible for apoptosis (programmed cell death) and act as signaling molecules that help regulate your optimal genetic expression. When your mitochondria are damaged or dysfunctional, not only will your energy reserves decrease, resulting in fatigue and brain fog, but you also become vulnerable to degenerative diseases such as cancer, heart disease, diabetes, and neurodegenerative decay.

Exercise Will Improve Your Immune Function

Aside from eating a varied, whole-food (ideally organic) diet and implementing time-restricted eating, exercise is a foundational health strategy that will strengthen your immune function.[17] According to research published in the March 19, 2020, issue of *Redox Biology*, exercising regularly may also help prevent acute respiratory distress syndrome (ARDS), which is very common in COVID-19.[18] Another way in which physical activity can help protect against COVID-19 is by combating immunosenescence, the decline in immune system function that typically occurs with aging.[19] Immunosenescence is believed to be one reason why the elderly are at such increased risk for viral infections in general and COVID-19 specifically.

What's more, exercising can get you outside in nature, which conveys mental health benefits as well as physical health benefits, as your body synthesizes vitamin D from direct exposure to sunlight, which contributes enormously to immune function. We'll dive deeper into the benefits of vitamin D for COVID-19 below and in chapter 7.

There are many types of exercise, but my favorite is a type of resistance training called blood flow restriction (BFR) training, which involves slightly restricting arterial flow and obstructing the venous return from the muscle back to the heart. This is done by applying bands to your arms or legs while exercising with very low weights at high repetition.

It's my favorite largely because it is a nearly perfect strategy for anyone over 50 or 60 to gain muscle mass with minimal risk of injury. You don't even need weights to practice this technique. This book doesn't allow us to go into more details, but there are well over 100 pages of instructions on bfr.mercola.com, as well as instructional videos. In my mind, BFR is one of the most powerful strategies to keep yourself healthy in the long term.

Another simple and inexpensive alternative is to use resistance bands— elastic rubber bands or ropes available in different shapes, sizes, and resistance levels. Most brands offer light, medium, and heavy bands that are adjustable, allowing you to be creative with your workouts.

Disease Prevention Through Stress Reduction

A healthy diet and level of physical activity aren't the only ways you can support your overall health and well-being, and thus how well your body responds to SARS-CoV-2 infection. How you manage stress is also extremely relevant.

Of course, stress levels are especially high for many during this pandemic. Even the conservative CDC recognizes that COVID-19 is escalating feelings of anxiety and stress.[20] When you're stressed, your immune system's ability to fight off infection is reduced.[21] Stress also promotes inflammation.[22]

Some of the effects of stress are direct. For example, the hormone cortisol, released during moments of stress, can suppress an effective immune response by lowering the number of infection-fighting lymphocytes circulating in your body. But effects can also be indirect, such as interfering with sleep, or prompting unhealthy behavioral coping strategies like snacking, drinking, and smoking.

Relaxation techniques are an important therapeutic strategy for stress-related diseases.[23] One randomized controlled trial concluded that those who exercised or meditated had fewer severe acute respiratory illnesses than those who did neither.[24]

Meditation, reading, listening to music, engaging in an absorbing hobby, and talking to friends—even if it is across the internet—can all help you relax, as can crossword puzzles, walking outside, and practicing yoga. Whatever engages you fully and takes you out of your head for a while counts as relaxation, so find what works for you.

Also don't discount the importance of simply turning off the news. Fear is often perpetuated by misinformation that feeds into panic. Make a decision to turn off negative news feeds or change your thoughts surrounding what you see and hear.

You may also want to consider trying Emotional Freedom Techniques (EFT), which can help you clear negative thought patterns in just a few minutes. You can find detailed instructions on my website, Mercola.com, by searching for "EFT."

Improving Your Immune Function Through Supplementation

There are a number of especially helpful supplements that can help protect you from COVID-19 and reduce your chances of experiencing more severe outcomes.

Vitamin D

I'm starting with vitamin D here on purpose, as evidence of the connection between low vitamin D levels and worse COVID-19 outcomes is overwhelming. In fact, aside from insulin resistance, vitamin D deficiency has emerged as a primary risk factor for severe COVID-19 infection and death. Higher vitamin D levels have even been shown to lower your risk of testing positive for the virus in the first place.

I created the website StopCOVIDCold.com, where you can find a 40-page document, complete with many illustrations and graphics and hundreds of references, that goes deep into the science of vitamin D. There is also a shorter version for the lay public. Additionally, StopCOVIDCold.com has a great two-minute test that you can take to find out your risk of developing COVID.

The largest observational study to date on vitamin D and COVID-19 was published in the journal *PLoS One*, September 17, 2020.[25] It looked at data for 191,779 American patients with a mean age of 50 who were tested for SARS-CoV-2 between March and June 2020 and had had their vitamin D tested sometime in the preceding 12 months. It found:

- 12.5 percent of patients who had a vitamin D level below 20 ng/ml (deficiency) tested positive for SARS-CoV-2.
- 8.1 percent of those who had a vitamin D level between 30 and 34 ng/ml (adequacy) tested positive for SARS-CoV-2.
- Only 5.9 percent of those who had an optimal vitamin D level of 55 ng/ml or higher tested positive for SARS-CoV-2.

Notably, the researchers concluded that people with a vitamin D level of at least 55 ng/mL (138 nmol/L) had a 47 percent lower SARS-CoV-2 positivity rate than those with a level below 20 ng/mL (50 nmol/L).

Vitamin D supplements are readily available and one of the least expensive supplements on the market. All things considered, vitamin D optimization is likely the easiest and most beneficial strategy that anyone can do to minimize their risk of COVID-19 and other infections, and can strengthen your immune system in a matter of a few weeks.

There's also a growing body of evidence showing vitamin D plays a crucial role in disease prevention and maintaining optimal health in general. Vitamin D affects nearly 3,000 of your 30,000 genes, which helps explain its influence. You also have vitamin D receptors located throughout your body.

According to one large-scale study, having optimal vitamin D levels can slash your risk of at least 16 different types of cancer, including pancreatic, lung, ovarian, prostate, and skin cancers. Vitamin D from sun exposure also radically decreases your risk of autoimmune diseases such as multiple sclerosis (MS) and type 1 diabetes, and helps prevent osteoporosis, which is a significant concern for women in particular.

Obtaining vitamin D through sun exposure is preferable over supplementation, as your skin is designed to produce vitamin D in response to the sun. Unfortunately, many are unable to get enough sun exposure due to their geography or work restrictions, and if this is the case, taking a vitamin D3 supplement is strongly recommended.

Vitamin D optimization is particularly important for dark-skinned individuals, as the darker your skin, the more sun exposure you need to raise your vitamin D level, and as a result the higher your likelihood of vitamin D deficiency. Increased skin pigmentation reduces the efficacy of UVB exposure because melanin functions as a natural sunblock.

If you're very dark-skinned, you may need to spend about 1.5 hours a day in the sun to have any noticeable effect. For many working adults and school-aged children, this simply isn't feasible. Light-skinned individuals, on the other hand, may need only 15 minutes of full sun exposure a day, which is far easier to achieve.

Still, they too will typically struggle to maintain ideal levels during the winter. During winter months at latitudes above 40 degrees north, little or no UVB radiation reaches the surface of the earth. That said, residence at low latitude does not guarantee adequate vitamin D levels, either, since social and cultural norms may limit your sun exposure.[26]

A healthy blood level of vitamin D is considered to be at least 40 ng/mL, with the recommended level being in the range of 40 to 60 ng/mL. However, for optimal health and COVID-19 prevention, the number you should aim for is between 60 and 80 ng/mL. Here are the key steps to raising your vitamin D levels:

1. **First, measure your vitamin D level**—One of the easiest and most cost-effective ways of measuring your vitamin D level is to participate in GrassrootsHealth's personalized nutrition project, which includes a vitamin D testing kit. Your doctor can also order a simple blood test from any of your local labs.

2. **Assess your individualized vitamin D dosage**—GrassrootsHealth's Vitamin D*Calculator is a helpful tool. To calculate how much vitamin D you may be getting from regular sun exposure in addition to your supplemental intake, use the DMinder app.[27]

3. **Retest in three to six months**—Lastly, you'll need to remeasure your vitamin D level in three to six months, to evaluate how your sun exposure and/or supplement dose is working for you. Adjust your dose up or down as needed, and retest again in another three to six months. Once you've determined the dose required to get you to an optimal level, you can test just once a year.

Important Points to Consider When Supplementing with Vitamin D

A few things to consider when taking supplemental vitamin D: Extensive studies by GrassrootsHealth of over 15,000 people have shown that if you aren't getting sun exposure, the typical adult requires 6,000 to 8,000 units of vitamin D per day. Children would require proportionally less.

Additionally, since more than half the population does not get enough magnesium, and far more are likely deficient, magnesium supplementation is recommended when taking vitamin D supplements. This is because magnesium helps activate vitamin D. On average, those who take vitamin D without supplemental magnesium need 146 percent more vitamin D per day to achieve a healthy blood level of 40 ng/ml (100 nmol/L), compared with those who take at least 400 mg of magnesium along with their vitamin D supplement.[28]

It's also important to increase your vitamin K_2 intake when taking high-dose supplemental vitamin D to avoid complications associated with excessive arterial calcification. Combined intake of both supplemental magnesium and vitamin K_2 has a greater effect on vitamin D levels than either individually. You need a whopping 244 percent more oral vitamin D if you're not concomitantly taking magnesium and vitamin K_2.[29]

Other Supplements to Consider

In addition to vitamin D, several other nutritional supplements can be useful for the prevention (and in some cases early treatment) of COVID-19. NAC, zinc, melatonin, vitamin C, quercetin, and B vitamins appear to be among the top choices. On page 107, I include a list of other nutrients known to improve immune function and combat viral illnesses.

N-acetylcysteine (NAC)

NAC is a precursor to reduced glutathione, which appears to play a crucial role in COVID-19. According to one literature analysis, glutathione deficiency may actually be associated with COVID-19 severity, leading the author to conclude that NAC may be useful both for its prevention and treatment.[30]

The idea that NAC can be helpful against viral infections is not new. Previous studies have found that it reduces viral replication of certain viruses, including the influenza virus.[31] In one such study, the number needed to treat (NNT) was 0.5, which means for every two people treated with NAC, one will be protected against symptomatic influenza.[32] That's significantly better than influenza vaccines, which have an NNV (number needed to vaccinate) of 71, meaning 71 people must be vaccinated to prevent a single case of confirmed influenza.[33] It's even better than vitamin D, which has an NNT of 33.[34]

Importantly, NAC has been shown to inhibit the damaging cascade associated with cytokine storms, which is a major cause of COVID-19 death. Studies have also demonstrated that NAC helps improve a variety of lung-related problems, including pneumonia and ARDS,[35] both of which are common characteristics of COVID-19.

Many COVID-19 patients also experience serious blood clots, and NAC not only counteracts hypercoagulation, as it has both anticoagulant and platelet-inhibiting properties,[36] but also breaks down blood clots once they've formed.[37] As noted in a paper published in the October 2020 issue of *Medical Hypotheses*: "We hypothesize that NAC could act as a potential therapeutic agent in the treatment of COVID-19 through a variety of potential mechanisms, including increasing glutathione, improving T cell response, and modulating inflammation."[38]

As of this writing, 11 studies involving NAC for COVID-19 are listed on ClinicalTrials.gov.[39] Ironically, just as we're starting to realize its benefits against this pandemic virus, the US Food and Drug Administration is suddenly cracking down on NAC, claiming it is excluded from the definition of a dietary supplement.

Zinc

Zinc plays a very important role in your immune system's ability to ward off viral infections. Zinc gluconate,[40] zinc acetate,[41] and zinc sulfate[42] have all been shown to reduce the severity and duration of viral infections such as the common cold. Zinc is a key ingredient in COVID-19 treatment protocols using hydroxychloroquine (HCQ). It's also a key component of the MATH+ protocol. You'll learn more about both of these in chapter 7.

Like vitamin D, zinc helps regulate your immune function[43]—and a combination of zinc with a zinc ionophore, like hydroxychloroquine or quercetin, was in 2010 shown to inhibit SARS coronavirus in vitro. In cell culture, it also blocked viral replication within minutes.[44] Importantly, zinc *deficiency* has been shown to impair immune function.[45]

As an early treatment for COVID-19 and other viral infections, take 7 mg to 15 mg of zinc four times a day, ideally on an empty stomach, or with a phytate-free food. This dose should not be taken long-term, however. Take only it until you recover from the illness. Getting at least 1 mg of copper from food and supplements for every 15 mg of zinc you take is also helpful, as zinc supplementation can backfire if you do not also maintain a healthy zinc-to-copper ratio.

Keep in mind there are many food sources of zinc, so a supplement may not be necessary. I eat about 12 ounces of ground bison or lamb a day, which provides 20 mg of zinc, so supplementation is not necessary for me. Following the general dietary advice shared in this chapter should ensure you get adequate zinc intake. But if you're just starting to eat a healthier diet, a supplement could be right for you.

Melatonin

Melatonin is a hormone synthesized in your pineal gland and many other organs.[46] While it is best known as a natural sleep regulator, it also boosts immune function in a variety of ways and helps quell inflammation. Melatonin may prevent SARS-CoV-2 infection by:

- Recharging glutathione (glutathione deficiency has been linked to COVID-19 severity).[47]
- Regulating blood pressure (a risk factor for severe COVID-19).
- Improving metabolic defects associated with diabetes and insulin resistance (risk factors for severe COVID-19) via inhibition of the renin-angiotensin system (RAS).
- Promoting synthesis of progenitor cells for macrophages and granulocytes, natural killer (NK) cells, and T helper cells (immune cells).
- Enhancing vitamin D signaling.

As a potent antioxidant,[48] it also has the rare ability to enter your mitochondria,[49] where it helps "prevent mitochondrial impairment, energy failure, and apoptosis of mitochondria damaged by oxidation."[50] In addition melatonin supports cardiovascular health—and as you now know, there are tight connections between COVID-19 risk and heart disease/hypertension[51]—and may even help prevent or improve autoimmune diseases, including type 1 diabetes.[52]

Not only may melatonin be an effective treatment against COVID-19 because of the benefits outlined above, but it also appears to protect against SARS-CoV-2 infection in the first place. In one study, patients who used melatonin supplements had, on average, a 28 percent lower risk of testing positive for SARS-CoV-2. Blacks who used melatonin were 52 percent less likely to test positive.[53]

While it's difficult to make melatonin dosage recommendations based on the limited evidence currently at hand, I recommend starting low, at 1 mg or less. Be sure to take melatonin at night, before bed. Rising melatonin levels is the reason you feel sleepy in the evening, so it's ill advised to take it in the morning or during the day, when your natural level is (and should be) low. If you happen to wake up in the middle of the night, especially if you're exposed to a light source, you could also take some then, to help you go back to sleep.

Melatonin is best taken sublingually, in the form of either a spray or a sublingual tablet. Sublingually, it can enter your bloodstream directly and doesn't have to go through the digestive tract. As a result, its effect will be felt more rapidly.

Keep in mind, however, that it makes little sense to take a supplement unless you're also seeking to optimize your body's natural production. In the case of melatonin, this includes making sure you get good sleep on a regular basis and a good dose of natural sunlight around midday to synchronize your circadian clock so that your body produces melatonin at the appropriate time (late evening). As the evening wears on and the sun sets, you'll want to avoid bright and all blue lighting, as blue light inhibits melatonin synthesis. Blue lighting is predominant in LED lights and fluorescent bulbs that are "cool white," which are best avoided.

Vitamin C

A number of studies have shown that vitamin C can be very helpful in the treatment of viral illnesses, sepsis, and ARDS,[54] all of which are applicable to COVID-19. Its basic properties include anti-inflammatory, immunomodulatory, antioxidant, antithrombotic, and antiviral activities. At high doses, it actually acts as an antiviral drug, actively inactivating viruses. Vitamin C also works synergistically with quercetin.[55]

In March 2020, Northwell Health, the largest hospital system in New York, reported vitamin C was being "widely used" against COVID-19 within its

23 hospitals, in conjunction with hydroxychloroquine and azithromycin (an antibiotic). A landmark literature review published in December 2020 also recommends the use of vitamin C as an adjunctive therapy for respiratory infections, sepsis, and COVID-19. As explained in this paper:

> *Vitamin C's antioxidant, anti-inflammatory and immunomodulating effects make it a potential therapeutic candidate, both for the prevention and amelioration of COVID-19 infection, and as an adjunctive therapy in the critical care of COVID-19 . . .*
>
> *The evidence to date indicates that oral vitamin C (2–8 g/day) may reduce the incidence and duration of respiratory infections and intravenous vitamin C (6–24 g/day) has been shown to reduce mortality, intensive care unit (ICU) and hospital stays, and time on mechanical ventilation for severe respiratory infections . . .*
>
> *Given the favorable safety profile and low cost of vitamin C, and the frequency of vitamin C deficiency in respiratory infections, it may be worthwhile testing patients' vitamin C status and treating them accordingly with intravenous administration within ICUs and oral administration in hospitalized persons with COVID-19.*[56]

The beneficial antiviral effects of vitamin C apply to both the innate and adaptive immune systems. When you have an infection, vitamin C improves your immune function in part by promoting the development and maturation of T lymphocytes, a type of white blood cell that is an essential part of your immune system. Phagocytes, immune cells that kill pathogenic microbes, are also able to take in oxidized vitamin C and regenerate it to ascorbic acid.

With regard to COVID-19 specifically, vitamin C:[57]

- Helps downregulate inflammatory cytokines, thereby reducing the risk of a cytokine storm. It also reduces inflammation through the activation of NF-κB and by increasing superoxide dismutase, catalase, and glutathione. Epigenetically, vitamin C regulates genes involved in the upregulation of antioxidant proteins and downregulation of proinflammatory cytokines.
- Protects your endothelium from oxidant injury.
- Helps repair damaged tissues.
- Upregulates expression of type I interferons, your primary antiviral defense mechanism, which SARS-CoV-2 downregulates.

- Eliminates ACE2 upregulation induced by IL-7. This is particularly noteworthy because the ACE2 receptor is the entry point for SARS-CoV-2 (the virus's spike protein binds to ACE2).
- Appears to be a powerful inhibitor of M^{pro}, a key protease (enzyme) in SARS-CoV-2 that activates viral nonstructural proteins.
- Regulates neutrophil extracellular trap formation (NETosis), a maladaptive response that results in tissue damage and organ failure.
- Enhances lung epithelial barrier function in an animal model of sepsis by promoting epigenetic and transcriptional expression of protein channels at the alveolar capillary membrane that regulate alveolar fluid clearance.
- Mediates the adrenocortical stress response, particularly in sepsis.

Vitamin C is a core component of the Front Line COVID-19 Critical Care working group's MATH+ protocol,[58] which will be reviewed in chapter 7. If used prophylactically, they recommend a dosage of 500 mg per day.[59]

Far higher dosages are required for the treatment of acute illness. Actually, when treating sepsis and/or COVID-19, the dosages needed are so high they generally require IV administration. To simulate IV administration levels if you're treating acute illness at home, you could take upward of 6 grams (6,000 mg) of liposomal vitamin C per hour. Doses above 20 grams per day of oral non-liposomal vitamin C typically results in loose stools. Using liposomal or IV vitamin C will allow you to take up to 100 grams (100,000 mg) a day without encountering such problems.

Keep in mind that, prophylactically, it is not recommended to take such high doses. In fact, I discourage people from taking mega doses of vitamin C on a regular basis if they're not actually sick—because in high doses it essentially works like a drug, and doing so could result in nutritional imbalances. So rather than taking it all the time, simply start mega-dosing at the first sign of symptoms of illness, and continue until symptoms recede. When you're well, you typically don't need more than 200 mg to 400 mg per day.

The only contraindication to high-dose vitamin C treatment is if you are glucose-6-phosphate dehydrogenase (G6PD) deficient, which is a genetic disorder.[60] G6PD is required for your body to produce NADPH, which is necessary to transfer reductive potential to keep antioxidants, such as vitamin C, functional.

Because your red blood cells do not contain any mitochondria, the only way they can provide reduced glutathione is through NADPH, and since G6PD eliminates this, it causes red blood cells to rupture due to inability to compensate for oxidative stress.

Fortunately, G6PD deficiency is relatively uncommon, and can be tested for. People of Mediterranean and African decent are at greater risk.

Quercetin

Quercetin, a powerful immune booster and broad-spectrum antiviral, was featured in a review of emerging COVID-19 research published in the *Integrative Medicine* journal in May 2020.[61]

Quercetin was initially found to provide broad-spectrum protection against SARS coronavirus in the aftermath of the SARS epidemic that broke out in 2003,[62] and evidence suggests it may be useful for the prevention and treatment of SARS-CoV-2 as well.

Quercetin's antiviral capacity has been attributed to five main mechanisms of action:

1. Inhibiting the virus's ability to infect cells by transporting zinc across cellular membranes.
2. Inhibiting replication of already infected cells.
3. Reducing infected cells' resistance to treatment with antiviral medication.
4. Inhibiting platelet aggregation—and many COVID-19 patients suffer abnormal blood clotting.
5. Promoting SIRT2, thereby inhibiting the NLRP3 inflammasome assembly involved with COVID-19 infection.

With regard to SARS-CoV-2 infection specifically, quercetin has been shown to:

- Inhibit the SARS-CoV-2 spike protein from interacting with human cells.[63]
- Inhibit SARS-CoV-2-related cytokine production.[64]
- Regulate the basic functional properties of immune cells and suppress inflammatory pathways and functions.[65]
- Act as a zinc ionophore—a compound that shuttles zinc into your cells.[66] This is one of the mechanisms that can account for the effectiveness seen with hydroxychloroquine, which is also a zinc ionophore.
- Boost interferon response to viruses, including SARS-CoV-2,[67] and inhibit the replication of RNA viruses.[68]
- Modulate the NLRP3 inflammasome, an immune system component involved in the uncontrolled release of proinflammatory cytokines that occurs during a cytokine storm.[69]

- Exert a direct antiviral activity against SARS-CoV.[70]
- Inhibit the SARS-CoV-2 main protease.[71]

Like vitamin C, quercetin is part of the MATH+ protocol. For prophylactic use, the MATH+ protocol recommends taking 250 mg to 500 mg of quercetin per day.[72]

B Vitamins

B vitamins can also influence several COVID-19-specific disease processes, including:[73]

- Viral replication and invasion.
- Adaptive immunity.
- Cytokine storm induction.
- Hypercoagulability.

A paper published in the February 2021 issue of the journal *Maturitas* details how each of the B vitamins can help manage various COVID-19 symptoms:[74]

Vitamin B_1 (thiamine)—Thiamine improves immune system function, protects cardiovascular health, inhibits inflammation, and aids in healthy antibody responses. Vitamin B_1 deficiency can result in an inadequate antibody response, thereby leading to more severe symptoms. There's also evidence suggesting B_1 may limit hypoxia.

Vitamin B_2 (riboflavin)—Riboflavin in combination with ultraviolet light has been shown to decrease the infectious titer of SARS-CoV-2 below the detectable limit in human blood, plasma, and platelet products.

Vitamin B_3 (niacin/nicotinamide)—Niacin is a building block of NAD and NADP, which are vital when combating inflammation. According to one review, niacin might actually be a crucial player in the COVID-19 disease process by boosting NAD+ and thwarting the cytokine storm and the downstream damage it causes. As noted in the abstract:

> *Definitive antiviral properties are evidenced for niacin, i.e., nicotinic acid (NA), as coronavirus disease 2019 (COVID-19) therapy for both disease recovery and prevention, to the level that reversal or progression of its pathology follows as an intrinsic function of NA supply. . . .*
>
> *The downstream inflammatory propagation of . . . SARS-CoV-2 infection is entirely prohibited or reversed upstream out of the body to expeditiously restore health with well-tolerated dynamic supplementation of sufficient NA (i.e., ~1–3 grams per day).*[75]

Aside from markedly decreasing proinflammatory cytokines, niacin has also been shown to:[76]

- Reduce the replication of a number of viruses, including vaccinia virus, human immunodeficiency virus, enteroviruses, and hepatitis B virus.
- Reduce neutrophil infiltration.
- Have anti-inflammatory effect in patients with ventilator-induced lung injury.
- Modulate bradykinin storms, responsible for some of the more unusual symptoms of COVID-19, including its bizarre effects on your cardio-vascular system.

Vitamin B$_5$ (pantothenic acid)—Vitamin B$_5$ aids in wound healing and reduces inflammation.

Vitamin B$_6$ (pyridoxal 5'-phosphate/pyridoxine)—Pyridoxal 5'-phosphate (PLP), the active form of vitamin B$_6$, is a co-factor in several inflammatory pathways. Vitamin B$_6$ deficiency is associated with dysregulated immune function. Inflammation increases the need for PLP, which can result in depletion. In COVID-19 patients with high levels of inflammation, B$_6$ deficiency may be a contributing factor. B$_6$ may also play an important role in preventing the hypercoagulation seen in some COVID-19 patients.

Vitamin B$_9$ (folate/folic acid)—Folate, the natural form of B$_9$ found in food, is required for the synthesis of DNA and protein in your adaptive immune response.

Folic acid, the synthetic form typically found in supplements, was recently found to inhibit furin, an enzyme associated with viral infections, thereby preventing the SARS-CoV-2 spike protein from binding to and gaining entry into your cells.[77] The research suggests folic acid may therefore be helpful during the early stages of COVID-19.[78]

Another recent paper found that folic acid has a strong and stable binding affinity against SARS-CoV-2. This, too, suggests it may be a suitable therapeutic against COVID-19.[79]

Vitamin B$_{12}$ (cobalamin)—B$_{12}$ is required for healthy synthesis of red blood cells and DNA. A deficiency in B$_{12}$ increases inflammation and oxidative stress by raising homocysteine levels. Your body can eliminate homocysteine naturally, provided you're getting enough B$_9$ (folate), B$_6$, and B$_{12}$.[80]

Hyperhomocysteinemia—a condition characterized by abnormally high levels of homocysteine—causes endothelial dysfunction, activates platelet and coagulation cascades, and decreases immune responses. B_{12} deficiency is also associated with certain respiratory disorders. Advancing age can diminish your body's ability to absorb B_{12} from food,[81] so the need for supplementation may increase as you get older.

As noted in one paper:

> *A recent study showed that methylcobalamin supplements have the potential to reduce COVID-19-related organ damage and symptoms. A clinical study conducted in Singapore showed that COVID-19 patients who were given vitamin B_{12} supplements (500 µg), vitamin D (1000 IU) and magnesium had reduced COVID-19 symptom severity and supplements significantly reduced the need for oxygen and intensive care support.*[82]

Other Helpful Nutritional Supplements

In a February 2020 article in the journal *Progress in Cardiovascular Diseases*, Mark McCarty of the Catalytic Longevity Foundation, and James DiNicolantonio, PharmD, a cardiovascular research scientist at Saint Luke's Mid America Heart Institute, reviewed a number of nutraceuticals that may be useful against RNA viruses such as influenza and SARS-CoV-2.[83] Several of them have already been reviewed above. Others include:

Elderberry extract—Known to shorten influenza duration by two to four days and reduce the severity of the flu. Provisional recommended daily dosage: 600–1,500 mg.

Spirulina—Reduces influenza infection severity and lowers influenza mortality in animal studies. Provisional recommended daily dosage: 15 grams.

Beta-glucan—Reduces the severity of influenza infection and lowers influenza mortality in animal studies. Provisional recommended daily dosage: 250–500 mg.

Glucosamine—Upregulates   mitochondrial   antiviral-signaling   protein, reduces influenza infection severity, and lowers influenza mortality in animal studies. Provisional recommended daily dosage: 3,000 mg or more.

Selenium—Selenium deficiency increases the viral mutation rate, thereby promoting the evolution of more pathogenic strains capable of evading your immune system. Provisional recommended daily dosage: 50–100 micrograms.

Lipoic acid—Helps boost type I interferon response, which is important for both your innate and adaptive immune systems. As explained in a 2014 paper, type I interferons:

... induce cell-intrinsic antimicrobial states in infected and neighboring cells that limit the spread of infectious agents, particularly viral pathogens. Second, they modulate innate immune responses in a balanced manner that promotes antigen presentation and natural killer cell functions while restraining pro-inflammatory pathways and cytokine production. Third, they activate the adaptive immune system, thus promoting the development of high-affinity antigen-specific T and B cell responses and immunological memory.[84]

Sulforaphane—Helps boost type I interferon response (see above).

Thiamine—While not included in the list provided by McCarty and DiNicolantonio, thiamine (vitamin B₁) helps regulate innate immunity and is an important component of the MATH+ protocol (see chapter 7). Like quercetin, it works synergistically with vitamin C. Thiamine deficiency has been implicated in severe infections, and shares many similarities with sepsis, a primary cause of COVID-19 mortality. Thiamine deficiency is also relatively common in critically ill patients in general.

Resveratrol—A 2005 study in the *Journal of Infectious Diseases* found that resveratrol has the power to inhibit the replication of influenza A virus, significantly improving survival in influenza-infected mice. According to the authors, resveratrol "acts by inhibiting a cellular, rather than a viral, function," which suggests it "could be a particularly valuable anti-influenza drug."[85]

Other Prevention Tactics

While diet, exercise, and stress reduction plus good sleep habits and supplementation are simple and effective lifestyle strategies to boost your immune system, reverse chronic disease, and prevent COVID-19, there are other specific COVID-19 prevention strategies you should be aware of as well.

Humidification

The combination of low temperatures and low humidity is an ideal environment for the spread of viral infections. This plays a prominent role in seasonal changes for viral infections such as influenza. Such may be the case for COVID-19 as well. Perhaps most important, low humidity can increase the ability of the coronavirus to spread between people.

Humidity is the concentration of water vapor in the air. This is an important and often overlooked variable in maintaining good health. During winter months, cold temperatures and indoor heating lead to drier air with low humidity.

Dry air with low humidity can increase feelings of being congested as your sinus membranes dry out and become irritated. The authors of one study found that high humidity contributed to nasal patency, the experience of breathing through clear nostrils.[86]

Low humidity can also contribute to dry, irritated eyes and may be a factor in increasing the evaporation of your tears. Colder temperatures and lower humidity also tend to dry out your skin.

Knowledge that humidity plays a role in the rate of respiratory infections is not new. In one study published over three decades ago, researchers found that maintaining mid-levels of humidity could help to lower the rate of respiratory infections and allergies.[87]

In a paper published in the *Journal of Global Health*, scientists reviewed the literature and proposed that humidity may not only reduce transmission of viral infections, but also play a role in your immune response.[88] They suggested the increase in viral infections during the winter months is a function of damage to the mucosal barrier by dry air. Within the mucous membranes are glycans, which are chemical structures that are bonded to most proteins. When pathogens enter the body, glycans are involved.

Mucins add another layer of protection. These glycosylated proteins found in the mucosal barriers are a decoy trap for viruses. Once trapped, viruses are then expelled from the airway. While these barriers are highly effective, they require proper hydration to maintain functionality.

When mucous membranes are exposed to dry air, their protective function is impaired. The results from an animal study demonstrated that raising relative humidity to 50 percent decreased mortality from flu infections. The researchers found animals that lived in dry air had a reduction in their mucociliary clearance and the ability to repair tissue. They were also more susceptible to disease.[89]

There are several ways to increase the humidity in your home to 40 to 60 percent. This is the level many experts believe helps moisturize your membranes and reduce the risk of infection.[90] Strategies you can use to help maintain the health of your nasal and sinus membranes include:

- Employ a vaporizer or room humidifier (see the caution below).
- Breathe in steam from a hot cup of tea or coffee.
- Boil water on your stove to boost humidity in the room.
- Place bowls of water around your home to improve humidity as they evaporate.

If you decide to use a room humidifier, be particularly careful to keep the humidity levels between 40 and 60 percent. Consistently high levels of humidity will increase the risk of mold growth, which can have a devastating effect on your health.

The warm, moist environment of a humidifier is an excellent breeding ground for bacteria and fungi, so you must clean your machine according to the manufacturer's instructions at least once every three days. The water in the reservoir should be changed daily.

The Best Long-Term Defense Is to Optimize Your Diet and Care for Your Body

It really didn't take long before it became apparent that the COVID-19 pandemic was illustrative of a far more widespread pandemic, namely that of insulin resistance and metabolic inflexibility.

All of the comorbidities that dramatically increase your COVID-19 risks (including your risk of symptomatic COVID-19 illness, hospitalization, and complications resulting in death) are rooted in insulin resistance. Remove the insulin resistance, along with vitamin D deficiency, and very few people—except for very old and frail individuals—would be at significant risk from SARS-CoV-2 infection.

So it is high time to start looking at how we can improve our metabolic health in general, and avoid insulin resistance and vitamin D deficiency in particular. A healthy population simply isn't going to be as vulnerable to infectious diseases like COVID-19.

If we want people to survive the next pandemic, whatever that might be, then improving public health has got to be the number-one priority going forward. Waiting for a drug cure or vaccine is a fool's game. Health care really needs to start emphasizing strategies known to improve overall health rather than throwing drugs at symptoms that don't address the underlying causes. Robust immune function is necessary to effectively combat COVID-19, and the same is true for all other infectious disease.

CHAPTER SEVEN

Pharmaceutical Failures in the COVID-19 Crisis

By Dr. Joseph Mercola

There is a long-standing history of corruption and fraud in the pharmaceutical industry. In a December 7, 2019, article in *The Lancet*, Dr. Patricia García—affiliate professor of global health at Cayetano Heredia University in Lima, Peru, and a former minister of health—points out that "Corruption is embedded in health systems."[1]

In it, she argues that dishonesty and fraud in the health care system as a whole, including its academic and research communities, is "one of the most important barriers to implementing universal health coverage," yet this corruption is rarely if ever discussed, let alone addressed in any meaningful way.

García writes:

> *Policy makers, researchers, and funders need to think about corruption as an important area of research in the same way we think about diseases. If we are really aiming to achieve the Sustainable Development Goals and ensure healthy lives for all, corruption in global health must no longer be an open secret.*
>
> *Corruption is an open secret known around the world that is systemic and spreading. Over two-thirds of countries are considered endemically corrupt according to Transparency International . . . Corruption in the health sector is more dangerous than in any other sector because it is literally deadly . . .*
>
> *It is estimated that, each year, corruption takes the lives of at least 140,000 children, worsens antimicrobial resistance, and undermines all of our efforts to control communicable and non-communicable diseases. Corruption is an ignored pandemic.*[2]

García summarizes the history of corruption, how it got started, and what allows it to spread. As a general commonsense rule, the less transparent a health system is, the more corrupt it becomes. This is precisely what happens when there is weak adherence to the rule of law. A lack of accountability mechanisms further invites corruption into the mix, resulting in a plummeting of health system performance, quality, and efficiency.

García also points out the economic cost of medical corruption:

> *It is estimated that the world spends more than US $7 trillion on health services, and that at least 10–25% of global spending is lost directly through corruption, representing hundreds of billions of dollars lost each year.*
>
> *These billions lost to corruption exceed WHO's estimations of the amount needed annually to fill the gap in assuring universal health coverage globally by 2030. However, the true cost of corruption for people is impossible to quantify because it can mean the difference between wellness and illness, and life and death.*

There's also plenty of evidence to suggest that scientific fraud launched the COVID-19 pandemic and is used to keep it going. Aside from fraudulent PCR testing and mislabeling positive tests as medical "cases," another instance of scientific malfeasance—without which this pandemic could not have been declared in the first place—was the World Health Organization's redefinition of *pandemic*.

The WHO's original definition of the term was: ". . . when a new influenza virus appears against which the human population has no immunity, resulting in several, simultaneous epidemics worldwide with enormous numbers of deaths and illness."[3]

The key portion of that definition is "enormous numbers of deaths and illness." This definition was changed in the month leading up to the 2009 swine flu pandemic. The change was a simple but substantial one: They merely removed the severity and high mortality criteria, leaving the definition of *pandemic* as "a worldwide epidemic of a disease."[4]

This switch in definition allowed the WHO to declare swine flu a pandemic after a mere 144 people had died from the infection, worldwide, and it's why COVID-19 is still promoted as a pandemic even though it has caused no excess mortality.[5] We now have plenty of data showing that the lethality of COVID-19 is on par with the seasonal flu.[6] It may be different in terms of symptoms and complications, but the actual lethality is about the same, and the absolute risk of death is equivalent to the risk of dying in a car accident.[7]

By removing the criteria of severe illness causing high morbidity, leaving geographically widespread infection as the only criterion for a pandemic, the WHO and technocratic leaders of the world were able to bamboozle the global population into giving up our lives and livelihoods. Had it not been for this move, COVID-19 would have been a non-event.

Perhaps even more egregious, in December 2020 the World Health Organization radically changed the definition of *herd immunity*. Clearly, this change is meant to pave the way for draconian mass vaccination campaigns, and in so doing, they're erasing the very foundation of immunology! As reported by the American Institute for Economic Research:

> *The World Health Organization, for reasons unknown, has suddenly changed its definition of a core conception of immunology: herd immunity . . . Herd immunity speaks directly, and with explanatory power, to the empirical observation that respiratory viruses are either widespread and mostly mild (common cold) or very severe and short-lived (SARS-CoV-1).*
>
> *The reason is that when a virus kills its host . . . the virus does not spread to others. The more this occurs, the less it spreads . . . When it happens to enough people . . . the virus loses its pandemic quality and becomes endemic, which is to say predictable and manageable . . .*
>
> *This is what one would call Virology/Immunology 101. It's what you read in every textbook. It's been taught in 9th grade cell biology for probably 80 years . . . And the discovery of this fascinating dynamic in cell biology is a major reason why public health became so smart in the 20th century. We kept calm. We managed viruses with medical professionals: doctor/patient relationships . . .*
>
> *Until one day, this strange institution called the World Health Organization . . . has suddenly decided to delete everything I just wrote from cell biology basics. It has literally changed the science in a Soviet-like way. It has removed with the delete key any mention of natural immunities from its website. It has taken the additional step of actually mischaracterizing the structure and functioning of vaccines.*[8]

As late as June 9, 2020, the World Health Organization's website described *herd immunity* as "the indirect protection from an infectious disease that happens when a population is immune either through vaccination or immunity developed through previous infection."

Then, in mid-November 2020, they updated the website, erasing any notion that humans have immune systems that protect them against disease naturally. Instead, *herd immunity*, according to the World Health Organization, is "a concept used for vaccination, in which a population can be protected from a certain virus if a threshold of vaccination is reached."

What's more, they claim, "Herd immunity is achieved by protecting people from a virus, not by exposing them to it." This is as backward as it gets. It's just plain *wrong*. As noted by the American Institute for Economic Research:

> *This change at WHO ignores and even wipes out 100 years of medical advances in virology, immunology, and epidemiology. It is thoroughly unscientific—shilling for the vaccine industry in exactly the way the conspiracy theorists say that WHO has been doing since the beginning of this pandemic.*
>
> *What's even stranger is the claim that a vaccine protects people from a virus rather than exposing them to it. What's amazing about this claim is that a vaccine works precisely by firing up the immune system through exposure . . . This has been known for centuries. There is simply no way for medical science completely to replace the human immune system. It can only game it via what used to be called inoculation.*[9]

Death by Modern Medicine

In her book *Death by Modern Medicine*, Dr. Carolyn Dean discusses how, in the past 100 years, the treatment of symptoms with drugs has dominated the practice of "health care."[10] The end result is a sickness industry that kills more people each year than most are ever aware of, as this information is simply buried by the media.

In 2000 Dr. Barbara Starfield published a landmark paper showing 225,000 Americans die each year from iatrogenic causes, meaning their death is caused by a physician's activity, manner, or therapy.[11] Ironically, Starfield herself died from a medical error over 10 years later, when she suffered a lethal reaction to an inappropriately prescribed antiplatelet drug.

Remarkably, matters have not improved one whit since then. A 2016 study published in the *BMJ* estimated that medical errors kill 250,000 Americans each year.[12] That's an annual increase of about 25,000 people from Starfield's estimates, and these numbers may still be vastly underestimated, as deaths occurring at home or in nursing homes were not included.

Indeed, when they included deaths related to diagnostic errors, errors of omission, and failure to follow guidelines, the number of preventable hospital deaths skyrocketed to 440,000 per year, which begins to hint at the true enormity of the problem. This was long before the mistreatment administered during the COVID-19 pandemic.

Conflicts of Interest Threaten Public Health

Conflict of interest is another pervasive problem that threatens the integrity and validity of most studies. Investigations assessing the prevalence of scientific fraud and/or its impact show that the problem is widespread and serious, to the point of making most of "science-based" medicine a genuine joke.

We've been repeatedly faced with study findings that are clearly tainted with industry bias. For example, a 2014 study funded by the American Beverage Association purported to have found that diet soda makes you lose more weight than drinking no soda at all—a finding that blatantly contradicts a large body of research demonstrating that artificial sweeteners disrupt your metabolism and lead to greater weight gain than sugar-sweetened beverages.[13]

Disturbingly, conflicts of interest are present at all levels, including our most prestigious public health agencies. While the US Centers for Disease Control and Prevention has long fostered the perception of independence, claiming it does not accept funding from special interests,[14] the agency has in fact made itself beholden to Big Pharma by accepting millions in corporate donations through its government-chartered foundation, the CDC Foundation, which funnels those contributions to the CDC after deducting a fee.[15]

Several watchdog groups—including the US Right to Know (USRTK), Public Citizen, Knowledge Ecology International, Liberty Coalition, and the Project on Government Oversight—filed a petition urging the CDC to cease making these false disclaimers.[16]

According to the petition, the CDC accepted $79.6 million from drug companies and commercial manufacturers between 2014 and 2018 alone. This is beyond unacceptable, as the CDC was created to be a public health watchdog, not a shill for corporations. It has tremendous influence within the medical community, and part of this influence hinges on the concept that it's free of industry bias and conflicts of interest.

While the drug industry is quick to claim that anyone questioning its integrity is part of a "war against science," the evidence of industry malfeasance is simply too great and too disturbing to ignore.

Vaccines are a primary profit driver for the drug industry.[17] Merck, which is just one of several vaccine makers, reported over $6.1 billion in sales of their childhood vaccines during the first three quarters of 2019 alone.[18]

A January 2020 vaccine market report states that the global vaccine market was worth $41.7 billion as of 2019, and is estimated to hit $58.4 billion by 2024.[19] One of the factors behind this rapid growth is "the rising focus on immunization." Anyone thinking this focus isn't manufactured by the drug industry itself is fooling themselves.

There is a cultural war and collusion between many industries and federal regulatory agencies that results in a suppression of the truth about vital health issues. If this suppression continues, we will progressively erode the private individual rights that our ancestors fought so hard to achieve.

Remdesivir—A COVID Treatment Scam

Drug companies are often portrayed as benevolent entities that pour billions of dollars into research so they can create new drugs and vaccinations for the greater good. However, they spend far more on marketing than they do research. According to a *New York Times* column on the antiviral drug remdesivir, biotech giant Gilead Sciences started distributing remdesivir on a compassionate-use basis in January 2020.[20]

That drug companies offer medications to patients in crisis is again deemed noble and altruistic. The *Times* column even noted, "Given the stakes involved, it seems perverse not to root for Gilead's success . . . there should be no Big-Pharma haters in pandemics."[21] The reality, however, is that the pharmaceutical industry develops drugs using our taxpayer money, and then turns around and sells them back to us at enormously inflated prices.

The actual "donation" to treat patients on a compassionate basis is virtually insignificant since the drug costs them very little. The positive press by publications like the *New York Times* provides them with a halo that allows them to convince clinicians to prescribe this expensive drug, which has never shown any established clinical benefits and has not been proven to reduce the potential for death in those with severe disease. As such, it's a perfect example of Big Pharma's emphasis of profit over people.

The long-awaited price for remdesivir was announced June 29, 2020, by Gilead Sciences. While the drug has demonstrated only questionable benefits, Daniel O'Day, chairman and CEO of Gilead Sciences, believes Gilead balanced corporate profits and public health when they settled on $520 per vial, which equates to $3,120 for the recommended five-day course of treatment (on the first day, a double dose is given).[22]

Meanwhile, the Institute for Clinical and Economic Review (ICER) released the calculated total cost of production, packaging, and a small profit margin on May 1, 2020. The cost was rounded to $10 per vial.[23] While the exorbitant price of remdesivir is partially based on the assumption that it will reduce the length of hospital stays by four days, some physicians, including Dr. George Ralls with Orlando Health, report that the drug actually *increases* the length of hospital stays. He told ABC News: "Once they start on this medication . . . they need it for five days, so they are in the hospital longer than they would have normally been. So that could be a reason why our inpatient numbers have ticked up a little."[24]

Remdesivir Studies Lack Positive Results

Although Gilead Sciences continues to move forward in its distribution of remdesivir, other scientific evidence has not supported its use. In one study, published in the *New England Journal of Medicine*,[25] the scientists changed the end point measurements for the study, moving all to secondary outcome measures except the number of days to recovery, which was the single primary outcome measure at the conclusion of the study.[26]

Although there were significant problems with the research design, and consequently the data, the release of the study generated enthusiasm and triggered immediate action across many countries, including the US, to the point that the US Food and Drug Administration issued an emergency-use authorization for remdesivir on May 1, 2020. This opened the door for compassionate use of the drug.[27]

However, a randomized, double-blind, placebo-controlled investigation into remdesivir proved it doesn't work. Two hundred thirty-seven patients in 10 hospitals were enrolled and randomly assigned to either a treatment group or a placebo group. The results showed remdesivir was not associated with statistically significant clinical benefits, and had to be stopped early because it was believed to have caused adverse events.[28]

In another paper, published in the *International Journal of Infectious Diseases*, scientists reported the outcomes for five of the first patients treated with remdesivir in France.[29] All of the patients had been admitted with severe pneumonia related to SARS-CoV-2 infection. Of the five, four experienced serious adverse events.

A randomized controlled study published in the May 16–22, 2020, issue of *The Lancet* also failed to find a clinical of benefit for remdesivir treatment.[30] Importantly, more than twice as many patients in the remdesivir group discontinued their treatment due to adverse effects than the control group (12 percent compared with 5 percent of those given a placebo).

And yet, at the time of this writing, even after over one year of treatments with no better data to back up its effectiveness, remdesivir is the only approved treatment by the FDA.[31]

The Treatment of Acute COVID-19

As you may have guessed, I don't believe remdesivir is the answer for COVID-19. Fortunately, there are now several treatment options that have demonstrated high levels of effectiveness and success, which I will review next, starting with the one I believe to be the most valuable.

Nebulized Hydrogen Peroxide—
The Most Effective Therapy for Acute COVID-19

Nebulized hydrogen peroxide, originally pioneered in the early 1990s by Dr. Charles Farr, is probably the single most effective intervention for those who have acute COVID-19. It's my favorite intervention for acute viral illnesses in general, and I strongly believe it would prevent the majority of people from dying from COVID-19 if used.

If you use the search engine on mercola.com to search for "nebulized hydrogen peroxide," you will find a very detailed explanation of why this therapy works and how to do it. Alternatively, an instructional video can be found on Bitchute.com, as YouTube has censored it.

In terms of mechanics, it's highly likely that the peroxide forms a very powerful signaling function that stimulates the immune system to defeat whatever viral threat it's exposed to. Your immune cells actually produce hydrogen peroxide. This is in part how they kills cells that have been infected with a virus. It appears that nebulized hydrogen peroxide merely enables your immune cells to perform their natural function more effectively.

In addition to being highly effective, it's inexpensive and has no side effects when used at the very low doses recommended (0.1 percent, which is 30 times less concentrated than regular drugstore 3 percent peroxide).

The key is to have your nebulizer already purchased and ready to go so that you can use it at the sign of first symptoms. You can also use it concomitant with vitamin C, as they likely have a powerful synergy and use different complementary mechanisms.

There are basically two types of nebulizers: small handheld devices that use AA batteries and devices that you plug into the wall. The ones you plug into the wall are far more effective, so be sure to use one of those. The PARI Trek S is my favorite and used to be available on Amazon but now requires a

Starting Peroxide Concentration	Hydrogen Peroxide	+	Water (Filtered)	=	Ending Peroxide Concentration
3%	¼ tsp	+	7¼ tsp	=	0.1%
12%	¼ tsp	+	5 ounces	=	0.1%
36%	¼ tsp	+	15 ounces	=	0.1%

Figure 7.1. Hydrogen Peroxide Dilution Schedule.

business account. So you can order it at justnebulizers.com and say Dr. Mercola recommended it, as the device requires a physician order. I don't receive any commissions for orders.

As for the hydrogen peroxide, since you are diluting it by 30 to 50 times (see figure 7.1), stabilizers are not likely to present a problem, but to be safe, your best bet is to use *food-grade* peroxide. Also, do not dilute it with plain water, as the lack of electrolytes in the water can damage your lungs if you nebulize it. Instead, use saline, or add a small amount of salt to the water to eliminate this risk.

You need about one teaspoon of salt in a pint of water or a half a teaspoon in an eight-ounce cup. This will create a physiological solution that will not harm your lungs when you inhale it. You could use regular table salt but ideally, use a healthy salt, such as Himalayan, Celtic, or Redmond salt.

The MATH+ and I-MASK+ Protocols

The MATH+ protocol, developed by the Front Line COVID-19 Critical Care working group (FLCCC), is one of the best, most effective, critical care protocols for COVID-19 to date. The initial MATH+ protocol was released in April 2020.[32] Since then, it has been updated several times to include quercetin and a number of optional nutrients and drugs.

In addition to the full clinical in-hospital critical care protocol (MATH+), there's also a protocol for prophylaxis and early outpatient treatment (I-MASK+), based on the drug ivermectin, a heartworm medication that has been shown to inhibit SARS-CoV-2 replication in vitro.[33] The MATH+ treatment grew out of Dr. Paul Marik's vitamin-C-based sepsis protocol, as he and other doctors noticed there were many similarities between sepsis and severe COVID-19 infection, in particular the out-of-control inflammatory cascade.

As these protocols continue to undergo revisions as more is learned, I recommend visiting the FLCCC website—covid19criticalcare.com—for the latest versions and dosages. At the time of the publication of this book, the I-MASK+ prophylaxis protocol includes the following drugs and supplements:[34]

- Ivermectin
- Vitamin D₃

- Vitamin C
- Quercetin

- Zinc
- Melatonin

The early outpatient protocol for mildly symptomatic patients is identical except for the dosages, and includes the addition of aspirin and oxygen saturation monitoring. The MATH+ hospital treatment protocol for COVID-19 (at the time of publication) includes:[35]

- Methylprednisolone (a steroid medication)
- Intravenous vitamin C
- Thiamine
- Heparin (an anticoagulant)
- Ivermectin

- Vitamin D
- Atorvastatin
- Melatonin
- Zinc
- Famotidine
- Therapeutic plasma exchange

In addition to these medications, the protocol calls for high-flow nasal oxygen to avoid mechanical ventilation. It's really crucial to avoid going on a ventilator, as it tends to worsen the condition for most patients. It can damage the lungs and is associated with a mortality rate approaching nearly 98 percent in some centers.[36]

It's important to note that while heparin is an important part of the protocol due to the clotting complications in the microvasculature of the lung, N-acetylcysteine (NAC), covered in chapter 6, may be a better choice. It has a better safety profile and is likely as effective. What's more, although many are afraid of steroids, they are a crucial component of COVID-19 treatment. In a short essay co-written by the entire FLCCC team, they point out that:

The FLCCC created the MATH+ protocol based on our physicians' insights into COVID-19 as a steroid-responsive disease. This treatment recommendation went against all the major national and international health care societies that had misinterpreted the medical literature, a body of published evidence, which, upon careful and deep review, actually supported the use of corticosteroids in prior pandemics . . .

Thousands of patients who became critically ill with COVID-19 and who were suffering from massive inflammation may have been saved if this safe and powerful anti-inflammatory medicine had been provided.[37]

Ivermectin

The inclusion of ivermectin in both the MATH+ and I-MASK+ protocols makes sense, as preliminary evidence seems to suggest it can be useful at all stages of SARS-CoV-2 infection. That said, its real strength really appears to be as a preventive approach.

On December 8, 2020, FLCCC president Dr. Pierre Kory, former professor of medicine at Aurora St. Luke's Medical Center in Milwaukee, Wisconsin, testified before the Senate Committee on Homeland Security and Governmental Affairs, where he reviewed the evidence supporting the use of the drug. As noted on the FLCCC website:

> *The data shows the ability of the drug Ivermectin to prevent COVID-19, to keep those with early symptoms from progressing to the hyper-inflammatory phase of the disease, and even to help critically ill patients recover.*
>
> *Dr. Kory testified that Ivermectin is effectively a "miracle drug" against COVID-19 and called upon the government's medical authorities—the NIH, CDC, and FDA—to urgently review the latest data and then issue guidelines for physicians, nurse-practitioners, and physician assistants to prescribe Ivermectin for COVID-19 . . .*
>
> *Numerous clinical studies—including peer-reviewed randomized controlled trials—showed large magnitude benefits of Ivermectin in prophylaxis, early treatment and also in late-stage disease. Taken together . . . dozens of clinical trials that have now emerged from around the world are substantial enough to reliably assess clinical efficacy.*
>
> *Data from 18 randomized controlled trials that included over 2,100 patients . . . demonstrated that Ivermectin produces faster viral clearance, faster time to hospital discharge, faster time to clinical recovery, and a 75% reduction in mortality rates.*[38]

While a 75 percent reduction in mortality is impressive enough, a WHO-sponsored review suggests ivermectin can reduce COVID-19 mortality by as much as 83 percent.[39] Like hydroxychloroquine, ivermectin is an antiparasitic drug with a well-documented safety profile and "proven, highly potent, antiviral and anti-inflammatory properties."[40] It's been on the market since 1981 and is on the World Health Organization's list of essential medicines.

It's also inexpensive, with a treatment course costing less than $2 in countries such as India and Bangladesh.[41] While the US FDA has not yet approved ivermectin for prevention of or treatment for SARS-CoV-2,[42] studies have shown ivermectin:[43]

- Inhibits replication of many viruses, including SARS-CoV-2 and seasonal influenza viruses. In my article "COVID-19: Antiparasitic Offers Treatment Hope," I review data showing a single dose of ivermectin killed 99.8 percent of SARS-CoV-2 in 48 hours.
- Inhibits inflammation through several pathways.
- Lowers viral load.
- Protects against organ damage.
- Prevents transmission of SARS-CoV-2 when taken before or after exposure; speeds recovery and lowers risk of hospitalization and death in COVID-19 patients.

On January 6, 2020, members of the FLCCC presented evidence to the National Institutes of Health COVID-19 Treatment Guidelines Panel, which is working to update NIH guidance.[44] One week later the National Institutes of Health updated their stand on use of the drug with a statement saying they would not recommend for or against it.[45] As noted by the FLCCC: "By no longer recommending against Ivermectin use, doctors should feel more open in prescribing Ivermectin as another therapeutic option for the treatment of COVID-19. This may clear its path towards FDA emergency use approval."[46]

Hydroxychloroquine: Game Changer or Deadly Treatment?

That fraudulent science has been used to promulgate and worsen the COVID-19 pandemic can clearly be seen in the early dismissal of hydroxychloroquine. While many doctors working on the front lines of the pandemic came out praising its effectiveness early on, the drug was quickly vilified as ineffective, unproven, or lethally dangerous.

In Spain, where hydroxychloroquine was used by 72 percent of doctors, it was rated "the most effective therapy" by 75 percent of them. The typical dose used by a majority of doctors was 400 milligrams per day.

French science-prize-winning microbiologist and infectious disease expert Didier Raoult, founder and director of the research hospital Institut Hospitalo-Universitaire Méditerranée Infection, reported[47] that a combination of hydroxychloroquine and azithromycin, administered immediately upon diagnosis, led to recovery and "virological cure"—nondetection of SARS-CoV-2[48] in nasal swabs—in 91.7 percent of patients.

According to Raoult, the drug combination "avoids worsening and clears virus persistence and contagiousness in most cases." No cardiac toxicity was observed using a dose of 200 mg three times a day for 10 days, along with 500 mg of azithromycin on Day 1 followed by 250 mg daily for the next four days. The risk of cardiac toxicity was ameliorated by carefully screening patients and performing serial EKGs.

According to Dr. Meryl Nass, the wildly divergent views on hydroxychloroquine appear to have little to do with its safety and effectiveness against COVID-19, and more to do with a concerted and coordinated effort to prevent its use. Indeed, there are several reasons why certain individuals and companies might not want an inexpensive generic drug to work against this pandemic illness—for reference, a 14-day supply costs just $2 to manufacture[49] and can retail for as little as $20.[50]

One of the most obvious reasons is that it might eliminate the need for a vaccine or other antiviral medication under development.[51] Hundreds of millions of dollars have already been invested, and vaccine makers are hoping for a payday in the billions, if not trillions, of dollars.

In the United States a number of doctors have tried to counteract the false propaganda against hydroxychloroquine, including family physician Dr. Vladimir Zelenko. He co-authored a study in which they found that treating COVID-19 patients "as early as possible after symptom onset" with zinc, low-dose hydroxychloroquine, and azithromycin "was associated with significantly less hospitalizations and five times less all-cause deaths."[52]

As noted by Zelenko, the real virus killer in this combination is actually the zinc. The hydroxychloroquine merely acts as a zinc transporter, allowing it to get into the cell. The antibiotic, meanwhile, helps prevent secondary infections.

Other proponents of hydroxychloroquine-based protocols include America's Frontline Doctors, a group of physicians who formed this coalition specifically to counter the false narrative that hydroxychloroquine is too dangerous to use for COVID-19. In addition to using the drug in hospitalized patients, they also stress that hydroxychloroquine in combination with zinc—just one 200 milligram tablet of hydroxychloroquine every other week with daily zinc—is an effective prophylactic that could be given to anyone at high risk of infection. Members of the group were quickly censored by social media platforms, and at least one doctor lost her job. Zelenko ended up being investigated by a Baltimore federal prosecutor.

Keep in mind that while hydroxychloroquine is a useful tool, it must be used very early in the course of the illness, ideally immediately after exposure, because it works by slowing down viral replication. It's also worth noting that in areas where hydroxychloroquine is hard to get a hold of, quercetin is likely a more effective and less expensive alternative, as its primary mechanism of action is identical to that of the drug, in addition to having many other anti-inflammatory benefits.

Both are zinc ionophores, meaning they shuttle zinc into the cell. As indicated by Zelenko, there's compelling evidence to suggest the primary benefit of this protocol comes from the zinc, which effectively inhibits viral replication. The problem is that zinc does not readily enter cells, which is why a zinc ionophore is needed.

Dr. Harvey A. Risch, a professor of epidemiology at Yale School of Public Health, has also tried getting the message out about hydroxychloroquine. In a July 23, 2020, *Newsweek* op-ed, he wrote:

> *I have authored over 300 peer-reviewed publications and currently hold senior positions on the editorial boards of several leading journals. I am usually accustomed to advocating for positions within the mainstream of medicine, so have been flummoxed to find that, in the midst of a crisis, I am fighting for a treatment that the data fully support but which, for reasons having nothing to do with a correct understanding of the science, has been pushed to the sidelines. As a result, tens of thousands of patients with COVID-19 are dying unnecessarily . . .*
>
> *I am referring, of course, to the medication hydroxychloroquine. When this inexpensive oral medication is given very early in the course of illness, before the virus has had time to multiply beyond control, it has shown to be highly effective, especially when given in combination with the antibiotics azithromycin or doxycycline and the nutritional supplement zinc.*[53]

Medical Technocracy Made the Pandemic Possible

The efforts to prevent medical professionals from using hydroxychloroquine is further evidence that the COVID-19 pandemic has an ulterior motive. If the medical establishment really wanted to save as many people as possible from this infection, wouldn't they embrace any and all things that work? The fact that they went out of their way to vilify a decades-old drug with an excellent safety profile shows we aren't dealing with a real medical establishment; we're dealing with medical technocracy. The censoring and manipulation of medical information are part and parcel of the social engineering part of this system.

The National Institutes of Health itself published research in 2005 showing chloroquine is a potent inhibitor of SARS coronavirus infection and spread, actually having both prophylactic and therapeutic benefits.[54] As the director of the National Institute of Allergy and Infectious Diseases (NIAID), which is a part of the NIH, since 1984, Anthony Fauci should have been well aware of these findings, yet he has, on multiple occasions, gone on record stating either that these drugs don't work, that there's insufficient evidence, or that the evidence is only anecdotal.

Now, in addition to treating COVID-19 patients and minimizing deaths, the related drug chloroquine has been shown to inhibit influenza A, and this may be yet another reason for the suppression of hydroxychloroquine.[55] If an inexpensive generic drug can prevent influenza, then what would we need seasonal influenza vaccines for?

In short, the drug poses a significant threat to the drug industry in more ways than one. It could also eliminate one of the most powerful leverages for geopolitical power that the technocrats have, namely biological terrorism. If we know how to treat and protect ourselves against designer viruses, their ability to keep us in line by keeping us in fear vanishes.

All of this helps explain the outright fraudulent studies published on hydroxychloroquine, which were then used as media fodder to frighten the public, all while positive studies were censored and suppressed. In one instance the authors pulled the data set out of thin air. They made it up. That study was ultimately retracted, but the bad publicity had already done its job. In other instances, they used doses *known* to be toxic.

While doctors reporting success with the drug are using standard doses around 200 mg per day for either a few days or maybe a couple of weeks, studies such as the Bill and Melinda Gates–funded[56] RECOVERY Trial used 2,400 mg of hydroxychloroquine during the first 24 hours—three to six times higher than the daily dosage recommended[57]—followed by 400 mg every 12 hours for 9 more days for a cumulative dose of 9,200 mg over 10 days.

Similarly, the Solidarity Trial, led by the World Health Organization, used 2,000 mg on the first day, and a cumulative dose of 8,800 mg over 10 days.[58] These doses are simply too high. More is not necessarily better. Too much, and guess what? You might kill the patient.

Hydroxychloroquine isn't the only potential COVID-19 remedy targeted by authorities. Just as data emerged showing the benefits of NAC on the infection, the US Food and Drug Administration suddenly started cracking down on NAC, claiming it is excluded from the definition of a dietary supplement.

While the agency has as of yet not taken action against NAC due to anything related to COVID-19—it has primarily targeted companies that market NAC as a remedy for hangovers—members of the Council for Responsible Nutrition have expressed concern the FDA may end up targeting it more widely. Hopefully, the FDA will *not* end up blocking access to NAC supplements in the same way hydroxychloroquine access has been stifled.

The Swiss Protocol—Quercetin and Zinc

Should you have concerns about hydroxychloroquine, or simply don't have access to it, quercetin is a viable, and perhaps even preferable, alternative. Like hydroxychloroquine, quercetin is a zinc ionophore, so it has the same mechanism of action as the drug—it improves zinc uptake by your cells.

So instead of taking hydroxychloroquine and zinc, you could take quercetin and zinc. These are also the core ingredients in the Swiss Protocol,[59] developed by the Swiss Coverage Analysis group, a nonpartisan, nonprofit analysis group that investigates "geopolitical propaganda in Swiss and worldwide media." It bases its reporting on published research, case studies, and "precise doctor testimonies." The complete protocol, recommended to be taken for five to seven days, also includes an anticoagulant, an antibiotic, and a mucolytic, plus hydroxychloroquine (although, again, quercetin has virtually identical actions and benefits).

It's best to take quercetin and zinc in the evening, right before bedtime and several hours after your last meal. The reason for this is that both quercetin and fasting are senolytic, meaning they selectively kill senescent cells—old, damaged zombie cells that accumulate during aging and accelerate inflammation damage. During sleep, you're effectively fasting for (hopefully) eight hours or so, which is why it's best to take the quercetin before bed to maximize these anti-aging benefits.

CHAPTER EIGHT

Successful Protocols Suppressed

By Dr. Joseph Mercola

Infectious diseases have been a serious threat to health for longer than humans have existed. Our ancestors historically relied on healthy immune systems to defeat them. In the last 150 years, advances in nutrition and sanitation have radically reduced the damage of these infections.

However, for the past 60 years the pharmaceutical industry has progressively increased its push to have the public believe that vaccinations are the way to prevent infectious disease. As noted in chapter 7, the World Health Organization has now gone so far as to redefine *herd immunity* to imply that vaccines are *required* to protect us from viral illnesses, completely erasing the very mention of the human immune system and the crucial role it plays.

The push for mandatory vaccine use radically accelerated with the implementation of the National Childhood Vaccine Injury Act of 1986, which granted drug makers partial liability protection for harm caused by vaccine products. The law was historic acknowledgment by the US government that federally licensed and recommended and state-mandated childhood vaccines can cause injury and death. It created a federal vaccine injury compensation program as an administrative alternative to a lawsuit for parents who did not want to go to court to sue drug companies or doctors.

Then, over a period of 30 years, the law was amended by Congress and by federal agencies through rule-making authority, gutting the law's informing, recording, reporting, and research provisions secured by parents in the legislation and making federal compensation almost impossible to obtain so fewer vaccine-injured people could get compensated.

In 2011 the US Supreme Court in a split decision in *Bruesewitz v. Wyeth*, with Justices Sonia Sotomayor and Ruth Bader Ginsburg dissenting, effectively removed all remaining liability from vaccine manufacturers for harm caused by vaccines in the US. From 2011 forward, vaccine manufacturers would not be liable for vaccine injuries and deaths, even if there was evidence the company could have made a vaccine less likely to cause harm.

Drug companies have been fined tens of billions of dollars in damages for side effects of the drugs they manufacture, so this complete liability protection for government-recommended and state-mandated vaccines is an important part of their financial success. With COVID-19, the liability protection has expanded even further and has completely shielded them from having to pay for vaccine injuries under the Public Readiness and Emergency Preparedness Act (PREP).

The first COVID-19 vaccines to be rolled out were experimental messenger RNA vaccines manufactured by Pfizer/BioNTech and Moderna, which were granted an emergency-use authorization by the US Food and Drug Administation (FDA) in December 2020 to be distributed in the US; they were also released in the U.K. and Canada. The "Sputnik" COVID-19 vaccine was distributed in Russia. While these new coronavirus vaccines have received a lot of positive press, there are several serious safety concerns that remain yet to be addressed.

Administering experimental vaccines to millions of people when there is only limited short-term safety data available because they have been fast-tracked to market is beyond reckless. Absolutely no long-term safety studies have been done to assess whether they might cause seizures, cancer, heart disease, allergies, and/or autoimmune diseases, all of which have been observed with other vaccines and were reported in earlier coronavirus vaccine trials on animals.

Animal studies were bypassed entirely for COVID-19 vaccines due to fast-tracking in the US government's Operation Warp Speed program launched in early 2020. As a result, millions of humans—with all sorts of underlying conditions that might render them more prone to vaccine reactions and permanent damage or even death—have now become the primary test subjects.

Researchers have been trying to develop a coronavirus vaccine since the severe acute respiratory syndrome (SARS-1) outbreak in 2002. None of them have succeeded, and many have demonstrated serious, sometimes fatal, side effects. It's also important to remember that mRNA vaccines have never before been licensed for use in humans. There is no data in humans studied over time that might give us an indication about what types of long-term effects from COVID-19 vaccines we can expect in years to come. To expect these experimental fast-tracked coronavirus vaccines to succeed when others that have been tested over far longer periods of time have failed miserably is pure folly.

While many hitched their hope for a "return to normal" and a feeling of safety to the rollout of COVID-19 vaccines, it didn't take long before reports of serious side effects started emerging and raised questions about whether their alleged benefits truly outweigh the potential harms. Independent researchers who have analyzed the available clinical trial data also point out that the effectiveness of these vaccines appears to be wildly exaggerated.

Vaccine Effectiveness Vastly Overstated

In early November 2020 Pfizer sent the stock market soaring when it announced that analysis of clinical trial data showed the efficacy of its vaccine was more than 90 percent. Soon after, an efficacy rate of 95 percent was announced.[1] Moderna boasted similar success with a 94.5 percent efficacy rating in its clinical trials.[2] However, the definition of *efficacy* is not being discussed.

If you read Pfizer's and Moderna's press releases and other clinical trial information, you'll see they've left out some really crucial information. For example:[3]

- They don't specify the cycle threshold used for the PCR tests they base their COVID-19 case count on, which is crucial for determining the accuracy of those tests.
- They don't mention anything about hospitalizations or deaths.
- There is no information about whether the vaccines prevent asymptomatic infection with and transmission of the SARS-CoV-2 virus; if the vaccine efficacy rate only prevents moderate to severe symptomatic disease and not infection and transmission, it will be impossible to achieve herd immunity using the vaccine.
- There is no indication about how long protection against moderate to severe symptomatic disease lasts. Some researchers suggest frequent booster doses will be required, perhaps every three to six months or annually.

Number Required to Vaccinate to Prevent One Case

In a letter to the editor published by the *BMJ*, Dr. Allan Cunningham, a retired pediatrician in New York, pointed out that Pfizer's effectiveness rating fails to tell the story in a way that people can actually understand, and went on to estimate the number needed to vaccinate for Pfizer's vaccine. This number gives you a much clearer picture of what you can expect (emphasis ours):

> *Specific data are not given but it is easy enough to approximate the numbers involved, based on the 94 cases in a trial that has enrolled about 40,000 subjects: 8 cases in a vaccine group of 20,000 and 86 cases in a placebo group of 20,000.*

This yields a COVID-19 attack rate of 0.0004 in the vaccine group and 0.0043 in the placebo group. Relative risk (RR) for vaccination = 0.093, which translates into a "vaccine effectiveness" of 90.7% [100(1 - 0.093)]. This sounds impressive, but the absolute risk reduction for an individual is only about 0.4% (0.0043 - 0.0004 = 0.0039).

The Number Needed to Vaccinate (NNTV) = 256 (1/0.0039), which means that to prevent just one COVID-19 case 256 individuals must get the vaccine; *the other 255 individuals derive no benefit, but are subject to vaccine adverse effects, whatever they may be and whenever we learn about them.*[4]

In an article published by the Mises Institute, Dr. Gilbert Berdine, associate professor of medicine at Texas Tech University Health Sciences Center, helps explain this statistical manipulation by performing the same calculation for the Moderna vaccine (emphasis ours):

The Pfizer study had 43,538 participants and was analyzed after 164 cases. So, roughly 150 out 21,750 participants (less than 0.7%) became PCR positive in the control group and about one-tenth that number in the vaccine group became PCR positive.

The Moderna trial had 30,000 participants. There were 95 "cases" in the 15,000 control participants (about 0.6%) and five "cases" in the 15,000 vaccine participants (about one-twentieth of 0.6%). The "efficacy" figures quoted in these announcements are odds ratios . . .

When the risks of an event are small, odds ratios can be misleading about absolute risk. A more meaningful measure of efficacy would be the number [needed] to vaccinate to prevent one hospitalization or one death. Those numbers are not available.

An estimate of the number [needed] to treat from the Moderna trial to prevent a single "case" would be 15,000 vaccinations to prevent 90 "cases" or 167 vaccinations per "case" prevented, which does not sound nearly as good as 94.5% effective.[5]

Another important data point shielded from the public is the absolute risk reduction provided by these vaccines. The drug companies are experts in confusing physicians and the public by conflating absolute and relative risks. They have previously done this in spades with statin drugs and made tens if not hundreds of billions in profits. In a November 26, 2020, *BMJ* article, Peter Doshi, associate editor of the journal, pointed out that while Pfizer claims its vaccine has a 95 percent efficacy rate, this is the *relative* risk reduction. The *absolute* risk reduction is actually less than 1 percent.[6]

In a later article Doshi presented yet additional concerns.[7] For starters he points out that Pfizer did not consistently confirm whether test subjects who showed symptoms of COVID-19 were actually PCR-positive. Instead, a large portion of them were simply marked as "suspected COVID-19." The problem with this is that the 95 percent efficacy rating is based on PCR-confirmed cases only.

Since the data show there are 20 times more suspected cases than confirmed cases, the relative risk reduction may actually be as low as 19 percent, Doshi said, which is far below the 50 percent efficacy required for use authorization by regulators. What's more, if suspected cases occurred in people who had false negative PCR test results, then the vaccine's efficacy would be even lower.

Yet another data point that might have a bearing on Pfizer's efficacy rate was the exclusion of 371 participants from its efficacy analysis due to "important protocol deviations on or prior to 7 days after Dose 2." Of those, 311 were from the vaccine group while only 60 were in the placebo group.

Why were five times as many in the vaccine group excluded from the efficacy analysis than in the placebo group? And what exactly were these "protocol deviations" that caused them to be excluded? This is called stacking the deck so the results can be manipulated in the desired direction to "prove" efficacy, when it is merely a statistical manipulation.

Will COVID-19 Vaccine Save Lives, Reduce Hospitalizations, or Prevent Transmission?

Doshi has also pointed out that current trials are not designed to tell us whether the vaccines will actually save lives. And if they don't, are they really worth the risks that might be involved? "What will it mean exactly when a vaccine is declared 'effective'?," he asks in his November 26 article. "To the public this seems fairly obvious. 'The primary goal of a COVID-19 vaccine is to keep people from getting very sick and dying,' a National Public Radio broadcast said bluntly . . . Yet the current phase III trials are not actually set up to prove either. None of the trials [were] designed to detect a reduction in any serious outcome such as hospital admissions, use of intensive care, or deaths."[8]

Nor do the trials tell us anything about the vaccine's ability to prevent asymptomatic infection and transmission, as this would require testing volunteers twice a week for long periods of time—a strategy that is "operationally untenable," according to Tal Zaks, chief medical officer at Moderna.[9]

Major Safety Questions Still Remain

Aside from the question of whether the COVID-19 vaccines work as adver-
tised, a number of safety questions also remain. And when it comes to safety,
it's important to understand that only a few thousand verified healthy volunteers were exposed to the actual vaccine. The real beta testers are those who are
lining up to take the vaccines when they first become available.

A rather extraordinarily long list of safety questions can be made, starting
with: "What effect will RNA vaccines have on DNA?" According to a January
29, 2020, Phys.org article, research has shown RNA does have a "direct effect
on DNA stability."[10]

Might COVID-19 vaccines disrupt genes, and if so, which ones? This could
be a rather crucial detail. As just one example, when genes that are important
for the chemical compound 6-methyladenine are eliminated, neurodegenera-
tion has been shown to occur in both mice and humans.[11]

Another safety question involves the lipid nanoparticles used in the vaccines. In 2017 Stat News discussed Moderna's challenges in developing an
mRNA-based drug for Crigler-Najjar syndrome, a condition that can lead to
jaundice, muscle degeneration, and brain damage:

> In order to protect mRNA molecules from the body's natural
> defenses, drug developers must wrap them in a protective casing.
> For Moderna, that meant putting its Crigler-Najjar therapy in
> nanoparticles made of lipids. And for its chemists, those nanoparti-
> cles created a daunting challenge: Dose too little, and you don't get
> enough enzyme to affect the disease; dose too much, and the drug is
> too toxic for patients.
>
> From the start, Moderna's scientists knew that using mRNA to spur
> protein production would be a tough task, so they scoured the medical
> literature for diseases that might be treated with just small amounts of
> additional protein. "And that list of diseases is very, very short," said the
> former employee . . .
>
> Crigler-Najjar was the lowest-hanging fruit. Yet Moderna could
> not make its therapy work . . . The safe dose was too weak, and repeat
> injections of a dose strong enough to be effective had troubling effects on
> the liver in animal studies.[12]

So are the lipid nanoparticles used in today's COVID-19 vaccines any safer
than the ones deemed too dangerous for human trials a few years ago? As we'll

review later on in this chapter, anaphylactic reactions emerged as one of the first widely occurring side effects, and this could potentially be related to these nanoparticles. Since the mRNA are rapidly degraded, they must be complexed with lipids or polymers.

The COVID-19 vaccines use PEGylated lipid nanoparticles, and polyethylene glycol (PEG) is known to cause anaphylaxis.[13] The risk of autoimmune challenges also looms large.

In his article Berdine points out that "colleagues are concerned about possible autoimmune side effects that may not appear for months after vaccination." It's worth noting that none of the trials included immunocompromised volunteers, so the effects of these vaccines on people with suppressed immune function is wholly unknown.

This is a significant problem, seeing how an estimated 14.7 million to 23.5 million Americans suffer from some form of autoimmune disease,[14] and these people are also at increased risk for COVID-19 complications and death. If the vaccine exacerbates autoimmune problems, the outcome could be devastating for an extraordinary number of people.

Vaccine-Induced Paradoxical Immune Reactions Could Spell Disaster

If previous trials of coronavirus vaccines are any indication, there is plenty to be worried about when it comes to the potential for serious side effects from COVID-19 vaccines. A frequent problem found in those studies was antibody-dependent immune enhancement—something we've known about since the 1960s. In a nutshell, this is when a viral vaccine renders you *more* prone to severe disease and death if subsequently you are infected with the virus.

As explained by James Odell, OMD, ND, L.Ac. in a December 28, 2020, Bioregulatory Medicine Institute article:

> Over a span of 18 years there have been numerous coronavirus vaccine animal studies conducted, which unfortunately demonstrated significant and serious side-effects. Either the animals were not completely protected, became severely ill with accelerated autoimmune conditions, or died.
>
> Animal side effects and deaths were primarily attributed to what is called Antibody-Dependent Enhancement (ADE) . . . Virus ADE is a biochemical mechanism in which virus-specific antibodies

(usually from a vaccine) promote the entry and/or the replication of another virus into white cells such as monocytes/macrophages and granulocytic cells.

This then modulates an overly strong immune response (abnormally enhances it) and induces chronic inflammation, lymphopenia, and/or a "cytokine storm," one or more of which have been reported to cause severe illness and even death. Essentially, ADE is a disease dissemination cycle causing individuals with secondary infection to be more immunologically upregulated than during their first infection (or prior vaccination) by a different strain.

ADE of disease is always a concern for the development of vaccines and antibody therapies because the mechanisms that underlie antibody protection against any virus has a theoretical potential to amplify the infection or trigger harmful immunopathology. ADE of the viral entry has been observed and its mechanism described for many viruses including coronaviruses.

Basically, it was shown that antibodies target one serotype of viruses but only sub neutralize another, leading to ADE of the latter exposed viruses . . . Because ADE has been demonstrated in animals, coronavirus vaccine research never progressed to human trials, at least not till the recent SARS coronavirus-2 fast-track campaign.[15]

The risk of antibody-dependent immune enhancement, also known as paradoxical immune enhancement (PIE), was highlighted in the paper "Informed Consent Disclosure to Vaccine Trial Subjects of Risk of COVID-19 Vaccine Worsening Clinical Disease," published in the *International Journal of Clinical Practice*, October 28, 2020. "COVID-19 vaccines designed to elicit neutralizing antibodies may sensitize vaccine recipients to more severe disease than if they were not vaccinated," the paper states, adding:

Vaccines for SARS, MERS and RSV have never been approved, and the data generated in the development and testing of these vaccines suggest a serious mechanistic concern: Vaccines designed empirically using the traditional approach (consisting of the unmodified or minimally modified coronavirus viral spike to elicit neutralizing antibodies), be they composed of protein, viral vector, DNA or RNA and irrespective of delivery method, may worsen COVID-19 disease via antibody-dependent enhancement (ADE).[16]

Previous Coronavirus Vaccine Tests
Were Flagged for Safety Risks

This risk, however, was not communicated to Pfizer and Moderna clinical trial participants. If one or more COVID-19 vaccines turn out to cause this type of immune enhancement, we could be looking at an avalanche of critical illnesses and deaths as people start being exposed to any number of mutated SARS-CoV-2 strains.

The saddest part is that this information was known, yet suppressed. In May 2020 I interviewed Robert F. Kennedy, Jr., about this very issue, at which time he provided the following story:

> Coronavirus vaccine development began after three SARS epidemics had broken out, starting in early 2002. The Chinese, the Americans, the Europeans all got together and said, "We need to develop a vaccine against coronavirus." Around 2012, they had about 30 vaccines that looked promising.
>
> They took the four best of those and manufactured the vaccines. They gave those vaccines to ferrets, which are the closest analogy when you're looking at lung infections in human beings.
>
> The ferrets had an extraordinarily good antibody response, and that is the metric by which FDA licenses vaccines. So they thought, "We hit the jackpot." All four of these vaccines worked like a charm. Then something terrible happened. Those ferrets were then exposed to the wild virus, developed inflammation in all their organs, their lungs stopped functioning.
>
> The scientists remembered that the same thing had happened in the 1960s when they tried to develop an RSV vaccine, which is an upper respiratory illness very similar to coronavirus. At the time, they did not test it on animals.
>
> They went right to human testing. They tested it on about 35 children, and the same thing happened. The children developed a champion antibody response, robust, durable. It looked perfect, and then the children were exposed to the wild virus and they all became sick. Two of them died. They abandoned the vaccine. It was a big embarrassment to FDA and NIH.
>
> Those scientists in 2012 remembered that, so, they looked closer and they realized that there are two kinds of antibodies being produced by the coronavirus. There are neutralizing antibodies, which are the kind you want, which fight the disease, and then there are binding antibodies.

The binding antibodies actually create a pathway for the disease in your body, and they trigger something called a paradoxical immune response or paradoxical immune enhancement. What that means is that it looks good until you get the disease, and then it makes the disease much, much worse. Coronavirus vaccines can be very dangerous, and that's why even our enemies, people who hate you and me—Peter Hotez, Paul Offit, Ian Lipkin—are all saying, "You got to be really, really careful with this vaccine."

Early Trials Raised Concerns About mRNA Vaccine Side Effects

Now that the first batches of COVID-19 vaccine have been rolled out, we're starting to see a number of worrying effects, and there was cause for concern from the very start of Moderna's phase 1 trials, when 80 percent of participants in the 100 microgram dose group suffered systemic side effects.[17]

After the second dose, 100 percent experienced side effects. Despite that, this was the dosage Moderna chose to move forward with into later-phase trials. (In its highest dosage group, which received 250 mcg, 100 percent of participants suffered side effects after the first dose.)

On May 20, 2020, Robert F. Kennedy, Jr., warned that "the clinical trial results could not be much worse." He wrote: "Moderna did not release its clinical trial study or raw data, but its press release, which was freighted with inconsistencies, acknowledged that three volunteers developed Grade 3 systemic events defined by the FDA as 'Preventing daily activity and requiring medical intervention.' . . . A vaccine with those reaction rates could cause grave injuries in 1.5 billion humans if administered to 'every person on earth.'"

To understand why mRNA COVID-19 vaccines are so disconcerting, you need to understand how they're designed to function. The Moderna and Pfizer vaccines both use messenger RNA (mRNA) technology to instruct your cells to make the SARS-CoV-2 spike protein. This is the glycoprotein that attaches to the ACE2 receptor of your cells, which allows the virus to actually infect you.

The idea behind these mRNA vaccines is that by creating the SARS-CoV-2 spike protein, your immune system will produce antibodies in response. What has not been factored into this treatment is how to shut off the production of these proteins once they aren't needed. What happens when you turn your body into a viral protein factory, thus keeping antibody production activated on a continual basis with no ability to shut down?

Furthermore, as mentioned in Kennedy's quote, there are two types of antibodies: binding antibodies and neutralizing antibodies. Binding antibodies are incapable of preventing viral infection. Instead they trigger an exaggerated immune response, as detailed above. In an early press release, Moderna noted that vaccine recipients had *binding antibodies* "at levels seen in blood samples from people who have recovered from COVID-19." At the time of that press release, data from 25 of 45 participants showed only this binding antibody result.

Meanwhile, neutralizing antibody data were available for only 8 of 45 participants, and the neutralizing antibodies are likely to be the more important, seeing how they are the ones that actually fight infection. Considering the problems caused by binding antibodies in previous coronavirus vaccine trials, these results triggered warning bells.

As noted by Robert Kennedy, Jr.:

> *Moderna did not explain why it reported positive antibody tests for only eight participants. These outcomes are particularly disappointing because the most hazardous hurdle for the inoculation is still ahead; challenging participants with wild COVID infection.*
>
> *Past attempts at developing COVID vaccines have always faltered at this stage as both humans and animals achieved robust antibody response then sickened and died when exposed to the wild virus.*[18]

Later-phase trials have yielded similarly high rates of side effects for Moderna and Pfizer alike. As noted by Doshi back in November 2020: "Moderna's press release states that 9% experienced grade 3 myalgia and 10% grade 3 fatigue; Pfizer's statement reported 3.8% experienced grade 3 fatigue and 2% grade 3 headache. Grade 3 adverse events are considered severe, defined as preventing daily activity. Mild and moderate severity reactions are bound to be far more common."[19]

On top of all this, while data are still limited, researchers at University of Pennsylvania and Duke University list a number of potential adverse effects from mRNA vaccines, including local and systemic inflammation, stimulation of autoreactive antibodies, autoimmunity, edema (swelling), and blood clots.[20]

Some of these effects, such as systemic inflammation and blood clots, resemble severe symptoms of COVID-19 itself. Might that be an indication that mRNA vaccines can indeed worsen COVID-19 infection and lead to paradoxical immune enhancement reactions similar to the ones that killed the coronavirus-immunized ferrets once they were exposed to the coronavirus?

Reported COVID-19 Vaccine Side Effects

The most concerning side effect reported in later-stage vaccine trials was transverse myelitis—inflammation of the spinal cord.[21] However, now that the Moderna and Pfizer vaccines have been given to tens of thousands of people with all sorts of underlying conditions, we're starting to see a much wider range of disturbing effects.

Within weeks of the vaccines becoming available (primarily to front-line health care workers and nursing home residents), reports of serious side effects started emerging in popular media and on social media networks. Among them:

- Persistent malaise[22] and extreme exhaustion.[23]
- Anaphylactic reactions.[24]
- Multisystem inflammatory syndrome.[25]
- Chronic seizures and convulsions.[26]
- Paralysis,[27] including cases of Bell's palsy.[28]
- At least 75 cases of sudden death (55 in the US and 20 in Norway), many occurring within hours or days.[29]

According to a report by the US Centers for Disease Control and Prevention, by December 18, 2020, 112,807 Americans had received their first dose of COVID-19 vaccine. Of those, 3,150 suffered one or more "health impact events," defined as being "unable to perform normal daily activities, unable to work, required care from doctor or health care professional." That gives us a side effect rate of 2.79 percent.[30]

Extrapolated to the total US population of 328.2 million, we may then expect more than 9,156,000 Americans to be injured by the vaccine if every single man, woman, and child is vaccinated. Extrapolated across the global population, the harm will be truly mind-boggling.

One suspected culprit in the allergic reactions people are experiencing is polyethylene glycol. The link appears valid enough that the CDC is warning people with known allergy to PEG or polysorbate to avoid all mRNA COVID-19 vaccines.[31]

COVID-19 Vaccine Trials Were Rigged

While vaccine makers insist that any vaccine reaching the market will have undergone rigorous testing, the design of the trial protocols clearly demonstrates the abandonment of virtually any attempt to confirm human safety.

The vaccines received a passing grade even if their efficacy for preventing infection was nonexistent. Preventing infection wasn't even a criterion for a successful COVID-19 vaccine. The only criterion of success was a reduction of moderate to severe COVID-19 symptoms, and even then the reduction required was minimal. In a September 2020 *Forbes* article, William Haseltine highlighted the questionable end points of these trials: "We all expect an effective vaccine to prevent serious illness if infected. Three of the vaccine protocols—Moderna, Pfizer, and AstraZeneca—do *not* require that their vaccine prevent serious disease only that they prevent moderate symptoms which may be as mild as cough, or headache."[32]

To get a "passing" grade in the limited interim analysis, a vaccine needed to show a 70 percent efficacy. However, this does not mean it will prevent infection in 7 of 10 people. As explained by Haseltine: "For Moderna, the initial interim analysis will be based on the results of infection of only 53 people. The judgment reached in interim analysis is dependent upon the difference in the number of people with symptoms . . . in the vaccinated group versus the unvaccinated group. Moderna's success margin is for 13 or less of those 53 to develop symptoms compared to 40 or more in their control group."

The other vaccine makers based their results on a similar protocol, where only a limited number of vaccinated participants are exposed to the virus to evaluate the extent of their moderate to severe COVID-19 symptoms.

As if that's not eyebrow-raising enough, the minimum qualification for a "case of COVID-19" amounts to just one positive PCR test and one or two mild symptoms, such as headache, fever, cough, or mild nausea. Basically, all they're doing is seeing if the COVID-19 vaccines minimize common cold symptoms.

There's no telling whether they will ultimately prevent hospitalizations and deaths. In fact, none of the trials included failure to prevent hospitalization or death as a measure of success. Johnson & Johnson's trial is the only one that requires at least five severe COVID-19 cases to be included in the interim analysis. Common sense dictates that if the vaccines cannot prevent or reduce infection and transmission, hospitalization, or death, then they cannot possibly end the pandemic.

Vaccinations Have Worsened Pandemic Illness in the Past

The idea that the COVID-19 vaccine might worsen illness is primarily based on the factors reviewed earlier in this chapter, such as the risk of antibody-dependent immune enhancement. But we can also look to previous vaccination campaigns. There are many studies showing that the seasonal influenza vaccine can actually increase your risk of pandemic influenza, for example.

Research raising serious questions about annual flu shots and their impact on pandemic viral illnesses include a 2010 review in *PLoS Medicine*, which found receiving the seasonal flu vaccine increased people's risk of getting sick with pandemic H1N1 swine flu, and resulted in more serious complications.[33]

People who received the trivalent influenza vaccine during the 2008–09 flu season were between 1.4 and 2.5 times more likely to get infected with pandemic H1N1 in the spring and summer of 2009 than those who did not get the seasonal flu vaccine. The findings were confirmed by the team in a study done on ferrets. *MedPage Today* quoted Dr. Danuta Skowronski, a Canadian influenza expert with the British Columbia Centre for Disease Control: "There may be a direct vaccine effect in which the seasonal vaccine induced some cross-reactive antibodies that recognized pandemic H1N1 virus, but those antibodies were at low levels and were not effective at neutralizing the virus. Instead of killing the new virus it actually may facilitate its entry into the cells."[34]

In all, five other observational studies conducted across several Canadian provinces found identical results. These findings also confirmed preliminary data from Canada and Hong Kong. As Professor Peter Collignon, an Australian infectious disease expert, told ABC News at the time: "We may be perversely setting ourselves up that if something really new and nasty comes along, that people who have been vaccinated may in fact be more susceptible compared to getting this natural infection."[35]

Does Flu Vaccination Increase Your Risk of COVID-19?

So what about SARS-CoV-2? Is there any evidence to suggest influenza vaccines might render people more susceptible to this pandemic virus, too? So far, no one has looked at SARS-CoV-2 specifically, but there are recent findings showing that seasonal flu shots can worsen coronavirus infections in general, and SARS-CoV-2 is one of seven different coronaviruses known to cause respiratory illness in humans.[36]

A study published in the January 10, 2020, issue of the journal *Vaccine* found people were more likely to get some form of coronavirus infection if they had been vaccinated against influenza. As noted in this study, titled "Influenza Vaccination and Respiratory Virus Interference Among Department of Defense Personnel During the 2017–2018 Influenza Season":

> *Receiving influenza vaccination may increase the risk of other respiratory viruses, a phenomenon known as virus interference. Test-negative study designs are often utilized to calculate influenza vaccine effectiveness.*

The virus interference phenomenon goes against the basic assumption of the test-negative vaccine effectiveness study that vaccination does not change the risk of infection with other respiratory illness, thus potentially biasing vaccine effectiveness results in the positive direction.[37]

While seasonal influenza vaccination did not raise the risk of all respiratory infections, it was in fact "significantly associated" with unspecified coronavirus (meaning it did not specifically mention SARS-CoV-2) and human metapneumovirus (hMPV). Those who had received a seasonal flu shot were 36 percent more likely to contract coronavirus infection and 51 percent more likely to contract hMPV infection than unvaccinated individuals.[38]

Looking at the symptoms list for hMPV is telling, as the main symptoms include fever, sore throat, and cough.[39] The elderly and immunocompromised are at heightened risk for severe hMPV illness, the symptoms of which include difficulty breathing and pneumonia. All of these symptoms also apply for SARS-CoV-2.

An Astonishing 1 in 40 Are Injured by Vaccines

We often hear that vaccine injuries occur at a rate of one in one million. This, however, is a gross underestimation. In a videotaped debate with lawyer Alan Dershowitz on the constitutionality of vaccine mandates, Robert F. Kennedy, Jr., discussed an investigation by the US Department of Health and Human Services Agency for Healthcare Research and Quality (AHRQ).[40]

They conducted a machine cluster analysis of health data collected from 376,452 individuals who received a total of 1.4 million doses of 45 vaccines. Of these doses, 35,570 vaccine reactions were identified, which means a more accurate estimate of vaccine damage would be 2.6 percent of all vaccinations. This means 1 in 40 people—not 1 in 1 million—are injured by vaccines, and a clinician who administers vaccines will have an average of 1.3 adverse vaccine events per month. As mentioned earlier in this chapter, based on early CDC data, we may be looking at a side effect rate of 2.79 percent for the COVID-19 vaccine. It's astonishingly close to the 2.6 percent found in this far larger cluster analysis.

That vaccines cause injuries is not a hypothetical. As noted by Kennedy, the reason vaccine manufacturers were given immunity in the first place was that they admitted vaccines are unavoidably unsafe and there's no way to make them 100 percent safe.

The National Vaccine Injury Compensation Program (VICP) created under the 1986 National Childhood Vaccine Injury Act has previously paid out over $4 billion to patients permanently damaged or killed by vaccines.

If that number wasn't bad enough, to add insult to injury, that's just a small portion of all the cases filed in the VICP—less than 1 percent of people who are injured ever get to court, due to the high bar set for proving causation. The risk of vaccine side effects and injuries is particularly troubling in light of the fact that vaccine manufacturers are indemnified against harm that occurs from the use of their federally recommended and state-mandated vaccines.

In chapter 2 I reviewed the devastating effects caused by the fast-tracked 2009 swine flu vaccine for the European market, Pandemrix, which a couple of years later was causally linked to skyrocketing cases of childhood narcolepsy. Now, in the midst of another controversial pandemic, we're facing an even greater public health threat. Kennedy (and other health experts) predicts the COVID-19 vaccine may become the greatest public health disaster in history. He says:

> You're going to see a lot of people dropping dead. The problem is, Anthony Fauci put $500 million of our [tax] dollars into that vaccine. He owns half the patents. He has five guys working for him [who are] entitled to collect royalties.
>
> So, you have a corrupt system, and now they've got a vaccine that is too big to fail. They're not saying this was a terrible, terrible mistake. They're saying, "We're going to order 2 million doses of this [vaccine]"... And, they have no liability... No medical product in the world would be able to go forward with a [safety] profile like Moderna has.[41]

Indeed, no one involved will be held accountable or face any repercussions, just as GlaxoSmithKline was not held accountable for the narcolepsy cases caused by Pandemrix. Instead, they will all continue to profit while an unsuspecting public will line up as guinea pigs for yet another dangerous vaccine.

Special Court Created Just for Those Injured or Killed by a COVID "Countermeasure"

Buried in the March 17, 2020, *Federal Register*—the daily journal of the US government—in a document titled, "Declaration Under the Public Readiness and Emergency Preparedness Act for Medical Countermeasures Against COVID-19," is language that establishes a new COVID-19 vaccine court—similar to the federal vaccine court that already exists for injuries and deaths caused by federally recommended vaccines for children and pregnant women.[42]

The US vaccine industry operates under a liability shield unlike any other in existence. If virtually any other existing product injures or kills a person, its manufacturer is held accountable in a civil court of law. With FDA-licensed and CDC-recommended vaccines, however, this is not the case.

Thirty-five years ago Congress created the federally operated Vaccine Injury Compensation Program. Through this, the US Court of Federal Claims in Washington, DC, handles contested vaccine injury and death cases in what has become known as vaccine court. When you sue for a vaccine injury, you're actually suing the US government, and payouts are paid for by the US public via a small fee tacked on to each vaccine sold.

The newly established COVID-19 vaccine court appears largely the same, except instead of focusing on injuries or deaths related to the recommended vaccines for children and pregnant women, it will be centered on those stemming from a new COVID-19 vaccine. Journalist Jon Rappoport highlighted the relevant section in this document, which includes compensation for covered "countermeasures" for COVID-19, such as a vaccine:

> *Countermeasures Injury Compensation Program . . . Section 319F-4 of the PHS Act, 42 USC 247d-6e, authorizes the Countermeasures Injury Compensation Program (CICP) to provide benefits to eligible individuals who sustain a serious physical injury or die as a direct result of the administration or use of a Covered [COVID] Countermeasure [for instance, a vaccine].*
>
> *Compensation under the CICP for an injury directly caused by a Covered Countermeasure is based on the requirements set forth in this Declaration, the administrative rules for the Program, and the statute. To show direct causation between a Covered Countermeasure and a serious physical injury, the statute requires "compelling, reliable, valid, medical and scientific evidence."*[43]

Compensation has been notoriously difficult to obtain from the existing vaccine court, and getting money from the CICP will likely be even more difficult, considering virtually all side effects are routinely dismissed as coincidental, and proving "direct causation" when we know virtually nothing about how mRNA vaccines affect human biology may be next to impossible.

Meanwhile, vaccine makers have nothing to lose by marketing their experimental shots, even if they cause serious injury and death. As Rappoport's tongue-in-cheek statement suggests:

> *"We know—and don't ask us how—that millions of you are going to get headaches. To prevent that, we're going to hit all of you on the head*

with a very heavy sledgehammer. If, ahem, a few of you happen to sustain an injury or die, we have a court where your relatives can try to get money out of us. By the way, in this court, we'll do everything we can to deny you money. Good luck." Yes, the government knows exactly what's coming when they approve a COVID vaccine. And now, so do you.[44]

Do We Really Need a COVID-19 Vaccine?

A large amount of data strongly suggests the COVID-19 vaccine may be completely unnecessary, which means the global population is being bamboozled into participating in a dangerous and unprecedented experiment for no good reason whatsoever. For example:

- COVID-19 mortality is extremely low outside of nursing homes—99.7 percent of people recover from COVID-19. If you're under 60 years of age, your chance of dying from seasonal influenza is greater than your chance of dying from COVID-19.[45]
- As covered in chapter 5, data clearly show that COVID-19 has not resulted in excess mortality, meaning the same number of people who die in any given year, on average, have died in this year of the pandemic.[46]
- As we'll explore in the next section, multiple studies suggest that immunity against SARS-CoV-2 infection is more widespread than suspected, thanks to cross-reactivity with other coronaviruses that cause the common cold.
- It is unclear whether asymptomatic people infected with SARS-CoV-2 are more or less likely to spread SARS-CoV-2. A study looking at PCR test data from nearly 10 million residents in Wuhan city found that not a single one of those who had been in close contact with an asymptomatic individual (someone who tested positive but had no symptoms) had been infected with the virus. In all instances, virus cultures from people who tested positive but had no symptoms also came up negative for live virus.[47]

Most Are Already Immune to SARS-CoV-2 Infection

It's important to realize you have two types of immunity. Your innate immune system is primed and ready to immediately attack foreign invaders at any moment and is your first line of defense. Your adaptive immune system, on the other hand, "remembers" previous exposure to a pathogen and mounts a delayed but more permanent long-term response when a previous encountered infection is recognized.[48]

Your adaptive immune system is further divided into two arms: humoral immunity (B cells) and cell-mediated immunity (T cells). The B cells and T cells are manufactured as needed from specialized stem cells.

If you have never been previously exposed to a disease but are given antibodies from someone who was and become ill and then recover, you can acquire humoral immunity against that disease. Your humoral immune system can also become activated if there's cross-reactivity with another similar pathogen. As you can see from the list below, in the case of COVID-19, evidence suggests exposure to other coronaviruses that cause the common cold can confer immunity against SARS-CoV-2.

Cell, June 2020—This study found that 70 percent of samples from patients who had recovered from mild cases of COVID-19 had resistance to SARS-CoV-2 on the T cell level. Importantly, 40 to 60 percent of people who had *not* been exposed to SARS-CoV-2 also had resistance to the virus on the T cell level.[49]

According to the authors, this suggests there's "cross-reactive T cell recognition between circulating 'common cold' coronaviruses and SARS-CoV-2." In other words, if you've recovered from a common cold caused by a particular coronavirus, your humoral immune system may activate when you encounter SARS-CoV-2, thus rendering you resistant to COVID-19.

Nature Immunology, September 2020—This German study, much like the *Cell* study above, found that "Cross-reactive SARS-CoV-2 peptides revealed pre-existing T cell responses in 81 percent of unexposed individuals and validated similarity with common cold coronaviruses, providing a functional basis for heterologous immunity in SARS-CoV-2 infection."[50]

The term *heterologous immunity* refers to immunity that develops against a given pathogen after you've been exposed to a non-identical pathogen. In other words, even among those who were unexposed, 81 percent were resistant or immune to SARS-CoV-2 infection.

The Lancet Microbe, September 2020—This study found that rhinovirus infection, responsible for the common cold, largely prevented concurrent influenza infection by triggering the production of natural antiviral interferon.[51]

The researchers speculate that the common cold virus could potentially help protect against SARS-CoV-2 infection as well. Interferon is part of your early immune response, and its protective effects last for at least five days, according to the researchers. Co-author Dr. Ellen Foxman told UPI:

> *This may explain why the flu season, in winter, generally occurs after the common cold season, in autumn, and why very few people have both*

viruses at the same time. Our results show that interactions between viruses can be an important driving force dictating how and when viruses spread through a population.

Since every virus is different, we still do not know how the common cold season will impact the spread of COVID-19, but we now know we should be looking out for these interactions.[52]

Nature, July 2020—This Singaporean study found that common colds caused by the betacoronaviruses OC43 and HKU1 might make you more resistant to SARS-CoV-2 infection, and that the resulting immunity could be long lasting. Patients who recovered from SARS infection back in 2003 still had T cell reactivity to the N protein of SARS-CoV now, 17 years later. These patients also had strong cross-reactivity to the N protein of SARS-CoV-2.

The authors suggest that if you've beaten a common cold caused by OC43 or HKU1 betacoronavirus in the past, you may have a 50/50 chance of having defensive T cells that can recognize and help defend against SARS-CoV-2.[53]

Cell, August 2020—This Swedish study found that exposed individuals, even if they tested negative for SARS-CoV-2 antibodies, still had SARS-CoV-2-specific memory T cells that may provide long-term immune protection against COVID-19.[54] As explained by the authors:

> *Importantly, SARS-CoV-2-specific T cells were detectable in antibody-seronegative exposed family members and convalescent individuals with a history of asymptomatic and mild COVID-19. Our collective dataset shows that SARS-CoV-2 elicits broadly directed and functionally replete memory T cell responses, suggesting that natural exposure or infection may prevent recurrent episodes of severe COVID-19.*[55]

Additional support for the idea that herd immunity may already have been achieved in most countries comes from statisticians working with mathematical models. For example, as early as June 2020, Professor Karl Friston, a statistician, claimed that immunity against SARS-CoV-2, globally, could be as high as 80 percent.[56]

Friston's model also effectively vaporizes claims that social distancing is necessary, because once sensible behaviors such as staying home when sick are entered into it, the positive effect of lockdown efforts on "flattening the curve" simply vanish. In all likelihood, the global lockdowns were completely unnecessary, and certainly should not continue.

There's also data showing that up to 80 percent of people tested at clinics have COVID-19 antibodies (meaning they're immune), and while rates may be lower among the general population, it's quite likely that herd immunity already exists among certain populations. In a survey of random households in Mumbai, up to 58 percent of residents in poor areas had antibodies, compared with up to 17 percent in the rest of the city.[57]

Now, if it's true that a majority already have some measure of immunity against COVID-19 due to previous exposure to other coronaviruses, then we've probably already reached the threshold for natural herd immunity, and vaccinating every human on the planet (or close to it) is completely unnecessary. What's more, the threshold for herd immunity may be far lower than previously suspected, which makes global inoculation even less of a necessity.

Herd Immunity Threshold for COVID-19 Could Be Under 10 Percent

Initial estimates by health officials were that 70 to 80 percent of the population would need to be immune before herd immunity would be achieved. Now more than a dozen scientists claim that the herd immunity threshold is likely below 50 percent.

Herd immunity is calculated using reproductive number, or R-naught (R_0), which is the estimated number of new infections that may occur from one infected person.[58] R_0 of below 1 (with R_1 meaning that one person who's infected is expected to infect one other person) indicates that cases are declining while R_0 above 1 suggests cases are on the rise.

It's far from an exact science, however, as a person's susceptibility to infection varies depending on many factors, including their health, age, and contacts within a community. The initial R_0 calculations for COVID-19's herd immunity threshold were based on assumptions that everyone has the same susceptibility and would be mixing randomly with others in the community.

"That doesn't happen in real life," Dr. Saad Omer, director of the Yale Institute for Global Health, told the New York Times.[59] "Herd immunity could vary from group to group, and subpopulation to subpopulation," or even zip code. When real-world scenarios are factored into the equation, the herd immunity threshold drops significantly, with some experts saying it could be as low as 10 to 20 percent.

Data from Stockholm County, Sweden, shows a herd immunity threshold of 17 percent,[60] while researchers from Oxford, Virginia Tech, and the Liverpool School of Tropical Medicine found that when individual variations in susceptibility and exposure are taken into account, the herd immunity threshold dips *below* 10 percent.[61]

As noted in an essay by Brown University professor Dr. Andrew Bostom:[62] "Separate HIT [herd immunity threshold] calculations of 9%,[63] 10–20%,[64] 17%,[65] and 43%[66]—each substantially below the dogmatically asserted value of ~70%[67]— have been reported by investigators from Tel-Aviv University, Oxford University, University College of London, and Stockholm University, respectively."

In another article that he wrote for *Conservative Review*, Bostom said:

> *Naturally acquired herd immunity to COVID-19 combined with earnest protection of the vulnerable elderly—especially nursing home and assisted living facility residents—is an eminently reasonable and practical alternative to the dubious panacea of mass compulsory vaccination against the virus.*
>
> *This strategy was successfully implemented in Malmo, Sweden, which had few COVID-19 deaths by assiduously protecting its elder care homes, while "schools remained open, residents carried on drinking in bars and cafes, and the doors of hairdressers and gyms were open throughout."*[68]

Adding support to Bostom's conclusion that naturally acquired herd immunity is a far better strategy than mandatory vaccination is Tom Britton, a mathematician at Stockholm University, who told the *New York Times* that since viral infections naturally target the most susceptible during the first wave, "immunity following a wave of infection is distributed more efficiently than with a vaccination campaign . . ."[69]

WHO Changes the Meaning of Herd Immunity

In June 2020, WHO's definition of herd immunity, posted on one of their COVID-19 Q&A pages, was in line with the widely accepted concept that has been the standard for infectious diseases for decades. Here's what it originally said, courtesy of the Internet Archive's Wayback machine:

> *Herd immunity is the indirect protection from an infectious disease that happens when a population is immune either through vaccination or immunity developed through previous infection.*[70]

It should be noted that "immunity developed through previous infection" is the way it's worked since humans have been alive. Apparently, according to WHO, that's no longer the case. In October 2020, here's their updated definition of herd immunity, which is now a "concept used for vaccination":

"Herd immunity", also known as "population immunity", is a concept used for vaccination, in which a population can be protected from a certain virus if a threshold of vaccination is reached.

Herd immunity is achieved by protecting people from a virus, not by exposing them to it.

Vaccines train our immune systems to create proteins that fight disease, known as "antibodies", just as would happen when we are exposed to a disease but—crucially—vaccines work without making us sick. Vaccinated people are protected from getting the disease in question and passing it on, breaking any chains of transmission. Visit our webpage on COVID-19 and vaccines for more detail.

With herd immunity, the vast majority of a population are vaccinated . . . lowering the . . . overall amount of virus able to spread in the whole population. As a result, not every single person needs to be vaccinated to be protected, which helps ensure vulnerable groups who cannot get vaccinated are kept safe. This is called herd immunity. . . .

The percentage of people who need to have antibodies in order to achieve herd immunity against a particular disease varies with each disease. For example, herd immunity against measles requires 95% of a population to be vaccinated. The remaining 5% will be protected by the fact that measles will not spread among those who are vaccinated. For polio, the threshold is about 80%.

Achieving herd immunity with safe and effective vaccines makes diseases rarer and saves lives.[71]

This perversion of science implies that the only way to achieve herd immunity is via vaccination, which is blatantly untrue. The startling implications for society, however, are that by putting out this false information, they're attempting to change our perception of what's true and not true, leaving people believing that they must artificially manipulate their immune systems as the only way to stay safe from infectious disease.

Many respected scientists are now calling for a herd immunity approach to the pandemic, meaning governments should allow people who are not at significant risk of serious COVID-19 illness to go back to normal life.

Tens of thousands of medical practitioners and scientists have signed the Great Barrington Declaration, which calls for "focused protection" rather than blanket lockdowns:

> *We know that vulnerability to death from COVID-19 is more than a thousand-fold higher in the old and infirm than the young. Indeed, for children, COVID-19 is less dangerous than many other harms, including influenza. As immunity builds in the population, the risk of infection to all—including the vulnerable—falls.*
>
> *We know that all populations will eventually reach herd immunity— i.e. the point at which the rate of new infections is stable—and that this can be assisted by (but is not dependent upon) a vaccine. Our goal should therefore be to minimize mortality and social harm until we reach herd immunity.*
>
> *The most compassionate approach that balances the risks and benefits of reaching herd immunity, is to allow those who are at minimal risk of death to live their lives normally to build up immunity to the virus through natural infection, while better protecting those who are at highest risk. We call this Focused Protection.*[72]

It's All Part of the Plan

There's been considerable global resistance to mandatory COVID-19 vaccination, but even if the vaccine ends up being "voluntary," refusing to take it may end up having severe implications for people who enjoy their freedom.

The Commons Project, the World Economic Forum, and the Rockefeller Foundation have joined forces to create the CommonPass, a digital "health passport" framework expected to be adopted by most if not all nations.[73] In other words, if you want to travel, you're going to have to roll up your sleeves and hope you're not one of the unlucky ones who end up with a permanent health problem from the vaccine. Just how voluntary is the vaccine if you're required to have it if you want to leave the country at any point during the rest of your life?

The groundwork for CommonPass was laid out in an April 21, 2020, white paper by the Rockefeller Foundation, and based on this paper, it's clear that proof of vaccination is part of a permanent surveillance and social control structure—one that severely limits personal liberty and freedom of choice across the board.[74]

There's absolutely no indication that proof of vaccination status will become obsolete once the COVID-19 pandemic is declared over, and the reason for this is because the pandemic is being used as a justification for the Great Reset, which will usher in a new system of technocracy that relies on digital surveillance and social engineering to control the population.

Proof of vaccination allows for the rollout of a highly invasive form of tracking that will undoubtedly expand with time. The tracking system proposed by the Rockefeller Foundation demands access to other medical data right from the get-go, which tells us the system will have any number of other uses besides tracking COVID-19 cases.

For years, I and others have warned that unless you get involved in protecting vaccine choice, even if and when it doesn't affect you personally, eventually it will indeed affect you and it'll be too late to do anything about it. We're now at that point. This affects everyone, not just teachers and health care workers. It affects all ages.

Any company can implement compulsory COVID-19 vaccination. No one is automatically excluded. Anyone could soon have to face the choice of vaccination or unemployment. Most schools are already saying they'll require students and staff to get inoculated against COVID-19. As reported by *National Geographic*, depending on where you live and the political philosophy of the majority of representatives in your state legislature, refusing the vaccine may also bar you from:[75]

- Obtaining a driver's license or passport.
- Attending a sports game or concert.
- Getting an education.
- Boarding a train or other public transportation.
- Entering a store, restaurant, bar, coffee shop, or nail salon.
- Booking an appointment with a doctor.
- Checking into a hospital for surgery.
- Visiting a family member in a nursing home.
- Obtaining private health insurance and Medicaid or Medicare.

There can be little doubt that the CommonPass is a cog in this Great Reset plan. It's the beginning stage of mass tracking and tracing, under the guise of keeping everyone safe from infectious disease. Rest assured, it will not be limited to COVID-19. The pandemic is just the justification for ushering in radical limitations on personal freedom and a massive increase in surveillance.

Blindly Trusting Big Pharma Could Be One of the Worst Mistakes of Your Life

The drug industry and government health officials expect you to blindly trust that they have developed a safe and effective COVID-19 vaccine, even though they eliminated well over six years of important testing, and despite the fact that no long-term safety assessments have actually been done. Drug companies have a long history of fraudulent and immoral practices and have paid tens of billions of dollars in fines for their crimes. The opioid epidemic is but one glaring example where company executives knew they were causing harm and chose to do it anyway. To say that trusting these convicted criminal organizations is a mistake would likely be one of the most profoundly serious understatements of the century. At this time we have no way to accurately predict what the consequences of injecting mRNA into your body will be. The good news is that, as we covered in chapters 6 and 7, there are loads of strategies to improve your immune system, and inexpensive, effective treatments should you come down with COVID-19. When you add that together with the fact that the lethality of COVID-19 is far lower than reported in the media, and the likelihood that widespread natural herd immunity already exists, the need for a vaccine seems remote indeed.

CHAPTER NINE

Take Back Control

By Ronnie Cummins

With some 2.6 billion people around the world in some kind of lockdown, we are conducting arguably the largest psychological experiment ever . . .

—Dr. Elke Van Hoof, World Economic Forum,
April 9, 2020[1]

We have allowed out-of-control politicians, tech giants, pandemic profiteers, operatives from the military-industrial complex, Big Pharma, medical mal-practitioners, large multinational corporations such as Amazon and Walmart, and a cabal of global health and economic elites to ruthlessly exploit us under the guise of a global pandemic.

These plunderers have utilized media censorship, shoddy science, manipulated statistics, fake news, and coercive government policies in order to ruthlessly expand their enormous power and wealth. The technocrats now have the power to monitor, censor, frighten, divide, and control the body politic as never before.

As exiled US whistleblower Edward Snowden warns us, "As authoritarianism spreads, as emergency laws proliferate . . . Do you truly believe that when the first wave, this second wave, the 16th wave of the coronavirus is a long-forgotten memory, that these capabilities will not be kept?"[2]

Shutting down the world over a respiratory virus will undoubtedly go down in history as the most destructive decision ever made by public health "experts," the World Health Organization and its technocratic allies. Unless you understand its true purpose, you'd probably label it irrational, but there's nothing irrational about it—from the technocrats' point of view.

The destruction—both moral and economic—is necessary for the Great Reset to occur. The technocratic elite need everything and everyone to fall apart in order to justify the implementation of their new system. Without widespread desperation, the population of the world would never agree to what they have planned. Even as evidence mounts showing that COVID-19 is hardly the deadly pandemic it's been made out to be, technocrats are grasping at straws to keep it going.

Case in point, mere days before Christmas 2020, U.K. prime minister Boris Johnson announced that there's a new, mutated, up-to-70-percent more infectious, strain of SARS-CoV-2 on the loose.[3] The threat from this mutated virus was deemed so concerning that another round of even stricter stay-at-home orders, business shutdowns, and travel bans was issued, just in time for the holidays.

This despite the fact that the new strain was reportedly identified in September 2020. Why all of a sudden was it deemed an emergency a full three months later—especially considering the research still had not been done to actually confirm that it was actually 70 percent more infectious than previous strains?

Carl Heneghan, professor of evidence-based medicine at Oxford University's Nuffield Department of Primary Care, told the *Daily Mail*, "I've been doing this job for 25 years and I can tell you [that you] can't establish a quantifiable number in such a short time frame. Every expert is saying it's too early to draw such an inference."[4]

The *New York Times* reported that the U.K. restrictions would likely remain in effect for months. Considering that these unscientific strategies didn't work the first or second time around, it strains believability to think they would work a third (or fourth or fifth) time, no matter how long they're implemented.

In fact, as noted by Matt Ridley in a *Telegraph* op-ed, viruses naturally weaken over time as more and more people are exposed, so by implementing tougher lockdowns, the virus primarily spreads among the sickest, which allows the most lethal strains to dominate.[5] In other words, by shutting everything down, the natural weakening of COVID-19 is prevented, which is the precise opposite of what we want.

Indeed, anyone who knows anything about the Great Reset agenda can now see that lockdowns have nothing to do with public health. They are mere smokescreens for the greatest transfer (if not theft) of wealth in the history of the world.

The biggest losers are low- and middle-income earners, especially private business owners, who have been absolutely decimated while big-box stores and multinational companies report record-breaking profits. As noted by Frank Clemente, executive director of Americans for Tax Fairness, "Never before has America seen such an accumulation of wealth in so few hands."[6]

We Must Turn the Tide

To end the madness, we need to understand the true origins, nature, virulence, prevention, and treatment of COVID-19. Armed with a proper understanding, we can then take appropriate action. The good news is that while COVID-19 indeed poses serious health risks to the elderly and those with comorbidities, we now know it poses very little risk to most people, especially children and young adults.

Another piece of good news is that elements in the mainstream media are finally starting to open up to the overwhelming evidence of a lab leak, with a comprehensive 12,000-word article in *New York* magazine by bioweapons historian Nicholson Baker being published in early January 2021.[7] Identifying the source of the virus is crucial if we are to prevent a pandemic like this from occurring in the future.

Beyond grasping the preponderance of evidence that SARS-CoV-2 was a lab leak, most people are also still unaware of the fact that miscalibrated, overly magnified PCR tests are vastly inflating the number of COVID-19 "cases," thereby causing people to live in fear and willingly accept authoritarian measures and restrictions on their freedoms.

Moreover, most people have no idea that 94 percent of the death certificates for COVID-19 victims routinely list a number of serious comorbidities leading to death, but that the CDC has ordered doctors to say that in all cases where the deceased has tested positive, or is suspected to be infected, COVID-19 is to be listed as the "primary" cause of death.

Very few understand that literally hundreds of thousands of cases of the common cold, influenza, pneumonia, and a number of acute respiratory illnesses are often caused or accompanied by coronaviruses every year, and that it is very difficult to diagnose, categorize, and separate these infections from COVID-19.

By conflating and combining all these different illnesses together, fearmongers and pandemic profiteers have been able to create the impression that there is a second (or in some cases third) wave of COVID-19 that threatens the lives of millions.

Similarly, most people do not realize that youth and moderately healthy people are a thousand times less likely to become symptomatic and seriously ill from COVID-19 and spread the disease to the most vulnerable; while the elderly, the frail, and the morbidly ill can be protected without the rest of us being locked down and possibly impoverished for the rest of our lives.

Speaking of the most vulnerable, we need to recognize that a large portion of the COVID-19 deaths—36 percent, as reported by the *New York Times*—were caused by the complete failure to prevent the spread in nursing homes.[8] Had infectious disease protocols been uniformly followed, the death toll would have been far lower.

The Ever-Shifting Goalpost Is a
Clue That Things Are Not as They Seem

Initially, the primary justification for the tyrannical governmental interventions was to slow the spread of the infection so that hospital resources would not be overwhelmed. Yet the goalpost has repeatedly been moved. Two-week lockdowns turned into months in many areas.

Eventually, we were told everything would go back to normal as soon as a vaccine became available. But once the vaccines started rolling out, the narrative changed again, and we're now told we'll still need masks, social distancing, and lockdowns well into 2021 or even 2022, even with a vaccine.

Very little makes any sense anymore—unless you look at it from the perspective we've tried to present to you here, namely that this pandemic has been used as a convenient cover story (and may even have been pre-planned) to facilitate and hide the transfer of wealth to unelected technocrats who control the pandemic narrative, while simultaneously justifying the erosion of your personal freedoms and civil liberties.

Panic has been sustained using a combination of wildly manipulated data and flawed tests. It's all a mirage. Take a deep look at the facts, and panic vanishes. You realize there's nothing to fear. Not really.

Aside from PCR testing data, there's no evidence of a lethal pandemic at all. While people have indeed died from COVID-19, there are no excess deaths due to it.[9] The total mortality for 2020 is normal. So unless we think we should shut down the world and stop living because people die from heart disease, diabetes, cancer, the flu, or anything else, then there's no reason to shut down the world because some people happen to die from COVID-19.

The Way Forward

As we've chronicled in this book, we don't yet know whether the recklessly engineered SARS-CoV-2 virus was *deliberately released* or whether it *accidentally escaped* from a negligently managed, accident-prone dual-use biodefense/bioweapons lab in Wuhan, China. We do know, however, that a powerful network of global elites, including Bill Gates, the World Economic Forum, Big Tech, the Rockefeller Foundation, and the Pentagon clearly anticipated what was coming, and then consciously took advantage of the crisis by seeding and nurturing panic to advance their economic, technocratic, totalitarian, anti-democratic agenda.

We also know that it is an existential imperative that we continue to expose the international gene engineers and scientists whose criminal negligence

brought on this disaster and put an end to the genetic engineering and weaponization of viruses and bacteria once and for all, so that nothing like this
pandemic ever happens again.

As we continue to gather more evidence that SARS-CoV-2 was lab-
engineered and that all of the global elite's misleading science, medical malpractice, and pandemic-mongering are being weaponized in a coordinated
and diabolical plan called the Great Reset, we must begin to unite a critical
mass of the educated, angry, and dispossessed.

As Arjun Walia of *Collective Evolution* points out, our most powerful rally-
ing cry is simply this: "Is this the world we truly want to create? Is this what
we are limited to creating, and if not, what holds us back?"[10]

Will we regretfully look back on 2020 as a dress rehearsal for the Great
Reset? Do we want to live in fear and/or guilt and wear a basically useless,
fear-inducing, socially-isolating mask for the rest of our lives?

Of course not.

Natural health and meditation advocate Dawson Church puts it well:
"We're in the middle of this mass contagion of fear, and it is depressing our
immune systems, rendering us less resilient, affecting us psychospiritually,
making us less able to cope. That's when we need a bigger dose of positivity,
joy and gratitude. We need to do that deliberately. That means meditation, it
means consuming positive media. It means not exposing yourself to needless
negative emotions."[11]

But if we want to stop the Great Reset that is being furthered by power-
intoxicated globalists, and instead build a world from the grassroots up that
is based upon peace and justice, tolerance, freedom, individual choice, privacy,
freedom of speech, religion, constitutional rights, and regenerative health,
food, farming, and land use, we must do more than just complain in private or
tweet about it to our followers.

Now is the time to get organized.

We need a new family-farm-based agricultural system that can provide
"food as medicine," organic and healthy food for all, while regenerating the
environment and biodiversity.

We need a new economic system that provides meaningful, socially and
environmentally responsible work and a decent standard of living for all who
are willing to work.

We need to object to and refuse any and all efforts to mandate COVID-19
vaccines. This includes rejecting the fake "choice" of voluntary vaccination in
the face of draconian restrictions for those who refuse to get it.

The Threat of Central Bank Digital Currencies

We also need sound money, be it in the form of physical cash or decentralized, block-chain-type digital currencies that protect our privacy and independence. The Great Reset brings with it a brand-new all-digital system that is not based on currency in the way we currently know it.

It's really a social control system, because by removing paper currency and replacing it with a central bank digital currency (CBDC), your ability to engage in transactions can be weaponized to destroy your privacy, surveil you, and prevent you from making purchases or even make a living.

Everything you buy and sell will be monitored, and punishment can be meted out if a transaction, your behavior, or even your thoughts are deemed undesirable by whatever "standards" that happen to be in vogue that week.

The transhumanist agenda is also part of this. Through the use of injections or some other means of getting biosensors into you, your actual physical body will be connected, literally, to the financial system. Transhumanism and technocracy fit hand in glove, and can best be described as a digital slavery system where you are monitored and controlled 24/7.

Decentralized-Everything Is the Way Forward

Perhaps most important of all, we need a decentralized government and internet where the threat of censorship is eliminated and free speech is assured. As just one example of many, anyone who questions pharmaceutical products on any of the social media platforms now faces the risk of being deplatformed. Many also find themselves booted from digital finance platforms such as PayPal at the same time, which proves the point I was trying to make in the previous section.

You, the individual, should have the most rights, because laws are best applied as specifically and locally as possible. The concentration of global and federal powers comes at the expense of your individual rights.

Mercola.com and a number of similar websites have even been labeled a multinational security threat by British and American intelligence agencies, which are collaborating to eliminate "anti-vaccine propaganda" from public discussion using sophisticated cyberwarfare tools.[12]

Ask yourself, does concern for public health really justify censoring and eliminating financial transaction capabilities of those who raise questions about vaccine safety and mandatory vaccination policies? The fact that they're trying to shut down all conversations about vaccines—using warfare tactics and economic blackmail, no less—suggests that the planned mass vaccination campaign has little if anything to do with keeping the public healthy and safe. It's about controlling the public and ensuring compliance.

The question is: *Why?*

The medical industry, and the vaccine industry in particular, have severe trust and credibility challenges that they themselves created and continue to grow with the help of Big Tech and national intelligence agencies, which are going to extreme lengths to prevent counternarratives from getting out.

Never before has the US government allowed this kind of blanket censorship of the public discourse. It would never be allowed if the government did it, but by delegating the censorship to private corporations it is allowed. It should be indisputable that censorship is anathema to a democratically run, free, and open society. While there may not be a benefit to allowing misinformation to be disseminated, the risks of censoring are simply too grave to be justifiable.

Censorship will never be applied just to the information you despise. It will be applied to any information that is threatening to the elite class that is attempting to further control us.

Big Tech censorship is actually even more insidious than government censorship, because it's far more opaque. At least if the government says it's going to censor certain kinds of expression, there's some level of transparency in how that's being done. Private tech companies, on the other hand, move the goalpost at will, and they're never entirely clear about who will be censored, for what, exactly, or how. What's more, there's no real process for appeal.

The problem we face now is that censorship fortifies power and is very difficult to end once it has taken hold. This in turn does not bode well for individual freedom or democracy as a whole. Censorship is a direct threat to both. With that in mind, the fact that US and U.K. intelligence agencies are getting involved in censoring tells us something important.

It tells us it's not really about protecting public health. It's about strengthening government control over the population. The fact that intelligence agencies view vaccine safety advocates as a national security threat also tells us that government is now in the business of *protecting private companies*, essentially blurring the line between the two.

If you criticize one you criticize the other. In short, if you impede or endanger the profitability of private companies, you are now viewed as a national security threat, and this falls squarely within the parameters of technocracy, in which government is dissolved and replaced with the unelected leaders of private enterprise.

The right and freedom to critique the government is a hallmark of democracy, so this state-sponsored war against truthful information is clear evidence of a radical turn toward technocratic totalitarianism.

Standing at a Crossroads—Which Way Will We Go?

The hour is late, but there is still time to turn things around. We stand at the crossroads—will we choose dictatorship, offered to us by our transhumanistic, technocratic "overlords," or freedom and democracy? I invite you to join us in the fight for our lives and the lives of future generations. Join us as we educate and organize for a healthy, equitable, and regenerative future.

By thinking and acting locally—buying local foods and products, and engaging in local politics and local organizing—we start to cut off the lifeblood of individuals and companies that are pushing us in the wrong direction. As David Klooz warns in his book, *The COVID-19 Conundrum*: "If the COVID-19 hoax does not convince you to divest from the politicians and the corporations they serve, including divesting from big-business' goods and services, nothing will. Special interests just beta-tested turning entire nations into virtual prisons. If people allow it this time, their ability to do it again and to an even greater and more disruptive degree is all but guaranteed . . ."[13]

Each one of our personal choices creates consumer-market-driven pressure that affects change from the bottom up. Forget top-down authoritative solutions. It always ends up biting us when we look to federal politicians to do something good for us, as lobbyists and lawyers—whose pockets are far deeper—are always working to make things better for the elites, not the general population.

Realize that federal and international agencies are captured by technocrats and oligarchs, so work within your community, and within yourself.

We likely can't stop the medical establishment from doing exactly what it wants to do—remain a slave to Big Pharma and treat symptoms rather than address fundamental root causes of disease. But you can opt out of these systems to keep yourself, your family, and your community healthy and resilient.

As detailed in chapter 6, simply optimizing your vitamin D can go a long way toward strengthening your immune system to ward off infectious disease, including COVID-19. Several other strategies are laid out in that chapter as well. The evidence for vitamin D in particular is so compelling that more than 100 doctors, scientists, and leading authorities have signed an open letter calling for increased use of vitamin D in the fight against COVID-19.[14]

"Research shows low vitamin D levels almost certainly promote COVID-19 infections, hospitalizations, and deaths. Given its safety, we call for immediate widespread increased vitamin D intakes," the letter states, adding:

> *Vitamin D modulates thousands of genes and many aspects of immune function, both innate and adaptive.*
>
> *The scientific evidence shows that:*
>
> - *Higher vitamin D blood levels are associated with lower rates of SARS-CoV-2 infection.*
> - *Higher D levels are associated with lower risk of a severe case (hospitalization, ICU, or death).*
> - *Intervention studies (including RCTs) indicate that vitamin D can be a very effective treatment.*
> - *Many papers reveal several biological mechanisms by which vitamin D influences COVID-19.*
> - *Causal inference modelling, Hill's criteria, the intervention studies & the biological mechanisms indicate that vitamin D's influence on COVID-19 is very likely causal, not just correlation.*[15]

Be the Light That Ends the 'Dark Winter'

While world leaders have been warning us of a "dark winter" with rising death tolls, hold on to hope. The way we end the pandemic fraud and stop the Great Reset is by shining the light of truth and creating greater transparency. If enough people end up understanding what's really going on and what the goal of this "Reset" is, leaders won't be able to implement it.

The technocratic elite need us all to passively acquiesce, because there are far more of us than there are of them. Again, that's what the pandemic measures are achieving. We're growing to accept work and travel restrictions. We're growing to accept government telling us where and how we can celebrate holidays, and with whom. All of this would have been unthinkable pre-2020. But we cannot let this acceptance continue to grow.

Slavery is the most profitable business in the history of the world, and with modern technology, complete control is now possible. Any and all rebellion can be quenched. Technology also allows a much smaller group of people to wield tremendous power over the masses.

That said, it's crucial to realize that *we* are actually financing and helping build the very control system that is meant to enslave us. We work for companies that are building the system. We buy products from them, and we allow them to collect data from us that they sell and utilize against us. If we stop buying their products and giving them our private data, they cannot build it. So stop giving them the means to enslave you. Stop being an enabler of your own doom. Instead, be part of the solution and shed light, share information, and seek out and develop alternatives to the control structure being erected all around us.

We've done this before. The organic movement, for example, was built by average people who decided to put their time and money into a food system that aligned with their basic values. As a result, we have options today when it comes to food. It's not all GMOs and fake food. If we want to live free, we now have to act on that wish, carefully reconstructing how we live and interact in order to minimize our contribution to the transhumanist technocratic control system.

Don't listen to nonsense about asymptomatic transmission, the PCR pandemic, and all the false stats used to scare you. Seek out the truth, take control of your health, and have frank and open discussions with family and friends to help show them the way out of fear, too.

Notes

Chapter One: How the Pandemic Plans Unfolded

1. Antonio Regolado, tweet, April 27, 2020, https://twitter.com/antonioregalado/status /1254916969712803840?lang=en.
2. Children's Health Defense Team, "An International Message of Hope for Humanity from RFK, Jr.," *Defender*, October 26, 2020, https://childrenshealthdefense.org/defender /message-of-hope-for-humanity.
3. Mary Holland, "What Can We Learn from a Pandemic 'Tabletop Exercise'?," Organic Consumers Association, March 25, 2020, https://www.organicconsumers.org/news/what -can-we-learn-pandemic-tabletop-exercise.
4. Martin Furmanski, "Laboratory Escapes and 'Self-Fulfilling Prophecy' Epidemics," Arms Control Center, February 17, 2014, https://armscontrolcenter.org/wp-content/uploads /2016/02/Escaped-Viruses-final-2-17-14-copy.pdf.
5. CDC COVID Data Tracker, "United States COVID-19 Cases and Deaths by State Reported to the CDC Since January 21, 2020," accessed January 20, 2021, https://covid .cdc.gov/covid-data-tracker/#cases_totaldeaths.
6. National Center for Health Statistics, "Weekly Updates by Select Demographic and Geographic Characteristics," CDC.gov, accessed August 26, 2020, https://www.cdc.gov /nchs/nvss/vsrr/covid_weekly/index.htm.
7. "Coronavirus Disease 2019: Older Adults," Centers for Disease Control and Prevention, updated September 11, 2020, https://www.cdc.gov/coronavirus/2019-ncov/need-extra -precautions/older-adults.html.
8. Craig Palosky, "COVID-19 Outbreaks in Long-Term Care Facilities Were Most Severe in the Early Months of the Pandemic, but Data Show Cases and Deaths in Such Facilities May Be on the Rise Again," KFF, September 1, 2020, https://www.kff.org/coronavirus -covid-19/press-release/covid-19-outbreaks-in-long-term-care-facilities-were-most-severe -in-the-early-months-of-the-pandemic-but-data-show-cases-and-deaths-in-such-facilities -may-be-on-the-rise-again.
9. Nurith Aizenman, "New Global Coronavirus Death Forecast Is Chilling—and Controversial," NPR online, September 4, 2020, https://www.npr.org/sections/goatsandsoda/2020/09/04 /909783162/new-global-coronavirus-death-forecast-is-chilling-and-controversial.
10. David M. Cutler and Lawrence H. Summers, "The COVID-19 Pandemic and the $16 Trillion Virus," *JAMA* 324, no. 15 (2020): 1495–96, https://www.doi.org/10.1001/jama.2020.19759.
11. Board of Governors of the Federal Reserve System, "Report on the Economic Well-Being of US Households in 2017," May 2018, https://www.federalreserve.gov/publications/files /2017-report-economic-well-being-us-households-201805.pdf.
12. Chuck Collins, "US Billionaire Wealth Surges Past $1 Trillion Since Beginning of Pandemic—Total Grows to $4 Trillion," Institute for Policy Studies, December 9, 2020, https:// ips-dc.org/u-s-billionaire-wealth-surges-past-1-trillion-since-beginning-of-pandemic/.

13. Naomi Klein, "Screen New Deal," *Intercept*, May 8, 2020, https://www.theintercept.com /2020/05/08/andrew-cuomo-eric-schmidt-coronavirus-tech-shock-doctrine.

14. Sainath Suryanarayanan, "Reading List: What Are the Origins of SARS-CoV-2? What Are the Risks of Gain-of-Function Research?," US Right to Know, updated October 13, 2020, https://usrtk.org/biohazards/origins-of-sars-cov-2-risks-of-gain-of-function -research-reading-list; "Attorney Dr. Reiner Fuellmich: The Corona Fraud Scandal Must Be Criminally Prosecuted for Crimes Against Humanity," FIAR News, October 9, 2020, https://news.fiar.me/2020/10/attorney-dr-reiner-fuellmich-the-corona-fraud-scandal -must-be-criminally-prosecuted-for-crimes-against-humanity; Organic Consumers Association Editors, "Covid-19: Right to Know," Organic Consumers Association, accessed November 18, 2020, https://organicconsumers.org/campaigns/covid-19.

15. Rowan Jacobsen, "Could COVID-19 Have Escaped from a Lab?," *Boston Magazine*, September 9, 2020, https://www.bostonmagazine.com/news/2020/09/09/alina-chan-broad -institute-coronavirus.

16. Ronnie Cummins and Alexis Baden-Mayer, "COVID-19: Reckless 'Gain of Function' Experiments Lie at the Root of the Pandemic," Organic Consumers Association, July 23, 2020, https://www.organicconsumers.org/blog/covid-19-reckless-gain-of-function -experiments-lie-at-the-root-of-the-pandemic.

17. Fred Guterl, Naveed Jamali, and Tom O'Connor, "The Controversial Experiments and Wuhan Lab Suspected of Starting the Coronavirus Pandemic," *Newsweek*, April 27, 2020, https://www.newsweek.com/controversial-wuhan-lab-experiments-that-may-have-started -coronavirus-pandemic-1500503.

18. Cambridge Working Group, "Cambridge Working Group Consensus Statement on the Creation of Potential Pandemic Pathogens (PPPs)," July 14, 2014, https://www.cambridge workinggroup.org.

19. "Scientists Outraged by Peter Daszak Leading Enquiry into Possible Covid Lab Leak," GM Watch, September 23, 2020, https://www.gmwatch.org/en/news/latest-news/19538.

20. Alexis Baden-Mayer, "Dr. Robert Kadlec: How the Czar of Biowarfare Funnels Billions to Friends in the Vaccine Industry," Organic Consumers Association, August 13, 2020, https://www.organicconsumers.org/blog/dr-robert-kadlec-how-czar-biowarfare-funnels -billions-friends-vaccine-industry.

21. Organic Consumers Association Editors, "'Gain of Function' Hall of Shame," Organic Consumers Association, October 1, 2020, https://www.organicconsumers.org/news/gain -of-function-hall-of-shame.

22. Dr. Joseph Mercola, "Can You Trust Bill Gates and the WHO with COVID-19 Pandemic Response?," Mercola.com, April 14, 2020, https://articles.mercola.com/sites/articles/archive /2020/04/14/world-health-organization-pandemic-planning.aspx.

23. Aksel Fridstøm, "The Evidence Which Suggests That This Is No Naturally Evolved Virus," *Minerva*, July 13, 2020, https://www.minervanett.no/angus-dalgleish-birger-sorensen -coronavirus/the-evidence-which-suggests-that-this-is-no-naturally-evolved-virus/362529; Bret Weinstein and Yuri Deigin, "Did Covid-19 Leak from a Lab?," *Bret Weinstein's Dark Horse Podcast*, June 8, 2020, https://www.youtube.com/watch?v=q5SRrsr-Iug.

24. Sam Husseini, "Did This Virus Come from a Lab? Maybe Not—But It Exposes the Threat of a Biowarfare Arms Race," Salon.com, April 24, 2020, https://www.salon.com/2020/04/24/did -this-virus-come-from-a-lab-maybe-not--but-it-exposes-the-threat-of-a-biowarfare-arms-race.

25. *Preventing a Biological Arms Race*, ed. Susan Wright (Cambridge, MA: MIT Press, 1990).

26. Lynn Klotz, "Human Error in High-Biocontainment Labs: A Likely Pandemic Threat," *Bulletin of the Atomic Scientists*, February 25, 2019, https://thebulletin.org/2019/02/human-error-in-high-biocontainment-labs-a-likely-pandemic-threat.

27. Dr. Joseph Mercola, "How COVID-19 Vaccine Can Destroy Your Immune System," Mercola.com, November 11, 2020, https://articles.mercola.com/sites/articles/archive/2020/11/11/coronavirus-antibody-dependent-enhancement.aspx.

28. Kristin Compton, "Big Pharma and Medical Device Manufacturers," Drugwatch, last modified September 21, 2020, https://www.drugwatch.com/manufacturers.

29. Leslie E. Sekerka and Lauren Benishek, "Thick as Thieves? Big Pharma Wields Its Power with the Help of Government Regulation," *Emory Corporate Governance and Accountability Review* 5, no. 2 (2018), https://law.emory.edu/ecgar/content/volume-5/issue-2/essays/thieves-pharma-power-help-government-regulation.html.

30. Dr. Joseph Mercola, "Swiss Protocol for COVID—Quercetin and Zinc," Mercola.com, August 20, 2020, https://articles.mercola.com/sites/articles/archive/2020/08/20/swiss-protocol-for-covid-quercetin-and-zinc.aspx.

31. Dr. Joseph Mercola, "How a False Hydroxychloroquine Narrative Was Created," Mercola.com, July 25, 2020, https://articles.mercola.com/sites/articles/archive/2020/07/15/hydroxychloroquine-for-coronavirus.aspx.

32. FLCCC Alliance, "FLCCC Summary of Clinical Trials Evidence for Ivermectin in COVID-19," January 11, 2021 (PDF), https://covid19criticalcare.com/wp-content/uploads/2020/12/One-Page-Summary-of-the-Clinical-Trials-Evidence-for-Ivermectin-in-COVID-19.pdf; Pierre Kory et al., "Review of the Emerging Evidence Demonstrating the Efficacy of Ivermectin in the Prophylaxis and Treatment of COVID-19," *Frontiers of Pharmacology*, provisionally accepted 2020, accessed January 21, 2021, https://doi.org/10.3389/fphar.2021.643369; "Ivermectin COVID-19 Early Treatment and Prophylaxis Studies," accessed January 20, 2021, https://c19Ivermectin.com.

33. Dr. Joseph Mercola, "Vitamin D Cuts SARS-Co-V-2 Infection Rate by Half," Mercola.com, September 28, 2020, https://articles.mercola.com/sites/articles/archive/2020/09/28/coronavirus-infection-rate-vitamin-d.aspx.

34. Dr. Joseph Mercola, "How Nebulized Peroxide Helps Against Respiratory Infection," Mercola.com, September 13, 2020, https://articles.mercola.com/sites/articles/archive/2020/09/13/how-to-nebulize-hydrogen-peroxide.aspx.

35. Dr. Joseph Mercola, "COVID-19 Critical Care," Mercola.com, May 29, 2020, https://articles.mercola.com/sites/articles/archive/2020/05/29/dr-paul-marik-critical-care.aspx.

36. Dr. Joseph Mercola, "Quercetin and Vitamin C: Synergistic Therapy for COVID-19," Mercola.com, August 24, 2020, https://articles.mercola.com/sites/articles/archive/2020/08/24/quercetin-and-vitamin-c-synergistic-effect.aspx.

37. Whitney Webb, "Operation Warp Speed Using CIA-Linked Contractor to Keep COVID-19 Vaccine Contracts Secret," Children's Health Defense, October 13, 2020, https://childrenshealthdefense.org/news/operation-warp-speed-cia-linked-contractor-covid-vaccine.

38. "Gates to a Global Empire," Navdanya International, October 14, 2020, https://navdanyainternational.org/bill-gates-philanthro-capitalist-empire-puts-the-future-of-our-planet-at-stake.

39. John Naughton, "'The Goal Is to Automate Us': Welcome to the Age of Surveillance Capitalism," *Guardian*, January 20, 2019, https://www.theguardian.com/technology/2019 /jan/20/shoshana-zuboff-age-of-surveillance-capitalism-google-facebook.

40. Dr. Joseph Mercola, "The Great Reset: What It Is and Why You Need to Know About It," Mercola.com, October 19, 2020, https://blogs.mercola.com/sites/vitalvotes/archive /2020/10/19/the-great-reset-what-it-is-and-why-you-need-to-know-about-it.aspx.

41. Alexis Baden-Mayer and Ronnie Cummins, "Gain-of-Function Ghouls: Sars-CoV-2 Isn't the Scariest Thing That Could Leak from a Lab," Organic Consumers Association, October 14, 2020, https://www.organicconsumers.org/blog/gain-function-ghouls-sars-cov-2-isnt -scariest-thing-could-leak-lab.

42. Frederik Stjernfelt and Anne Mette Lauritzen, *Your Post Has Been Removed* (New York: Springer, 2020).

43. "Truth to Power," Organic Consumers Association, accessed November 20, 2020, https://www.organicconsumers.org/newsletter/scientist-isnt-afraid-speak-truth-power /truth-power.

44. Dr. Joseph Mercola, "The Real Danger of Electronic Devices and EMFs," Mercola.com, September 24, 2017, https://articles.mercola.com/sites/articles/archive/2017/09/24 /electronic-devices-emf-dangers.aspx.

45. Natasha Anderson and Nexstar Media Wire, "New CDC Report Shows 94% of COVID-19 Deaths in US Had Contributing Conditions," WFLA, August 30, 2020, https://www .wfla.com/community/health/coronavirus/new-cdc-report-shows-94-of-covid-19-deaths -in-us-had-underlying-medical-conditions.

46. Katherine J. Wu, "Studies Begin to Untangle Obesity's Role in Covid-19," *New York Times*, September 29, 2020, updated October 14, 2020, https://www.nytimes.com/2020/09/29 /health/covid-obesity.html.

47. Barry M. Popkin et al., "Individuals with Obesity and COVID-19: A Global Perspective on the Epidemiology and Biological Relationships," *Obesity* 21, no. 11 (2020): e13128, https://doi.org/10.1111/obr.13128.

48. Shemra Rizzo et al., "Descriptive Epidemiology of 16,780 Hospitalized COVID-19 Patients in the United States," medRxiv preprint, 2020, https://doi.org/10.1101/2020.07.17.20156265.

49. Ronnie Cummins, "Genetic Engineering, Bioweapons, Junk Food and Chronic Disease: Hidden Drivers of COVID-19," Organic Consumers Association, September 30, 2020, https://www.organicconsumers.org/blog/genetic-engineering-bioweapons-junk-food -and-chronic-disease-hidden-drivers-covid-19.

50. Dr. Joseph Mercola, "Global Uprising Underway," Mercola.com, September 16, 2020, https://articles.mercola.com/sites/articles/archive/2020/09/16/global-uprising.aspx; Ronnie Cummins, *Grassroots Rising* (White River Junction, VT: Chelsea Green, 2020).

51. "US Found to Be Unhealthiest Among 17 Affluent Countries," *American Medical News*, January 21, 2013, https://amednews.com/article/20130121/health/130129983/4.

52. Andrew Hutchinson, "YouTube Ramps Up Action to Remove Covid-19 Misinformation," *Social Media Today*, April 23, 2020, https://www.socialmediatoday.com/news/youtube -ramps-up-action-to-remove-covid-19-misinformation/576577.

53. Dr. Joseph Mercola, "Oneness vs. the 1%," Mercola.com, November 1, 2020, https://articles .mercola.com/sites/articles/archive/2020/10/18/vandana-shiva-oneness-versus-the-1.aspx.

54. Children's Health Defense Team, "An International Message of Hope for Humanity."

Chapter Two: Lab Leak or Natural Origin?

1. *Washington Post* Editorial Board, "Opinion: The Coronavirus's Origins Are Still a Mystery. We Need a Full Investigation," *Washington Post*, November 20, 2020, https://www.washingtonpost .com/opinions/global-opinions/the-coronaviruss-origins-are-still-a-mystery-we-need-a-full -investigation/2020/11/13/cbf4390e-2450-11eb-8672-c281c7a2c96e_story.html?mc_cid =1f31114972&mc_eid=9723e894e5; David A. Relman, "Opinion: To Stop the Next Pandemic, We Need to Unravel the Origins of COVID-19," *Proceedings of the National Academy of Sciences* 117, no. 47 (November 2020), 29246–48; https://doi.org/10.1073/pnas.2021133117.

2. Lynn C. Klotz, "The Biological Weapons Convention Protocol Should Be Revisited," *Bulletin of the Atomic Scientists*, November 15, 2019, https://thebulletin.org/2019/11/the-biological -weapons-convention-protocol-should-be-revisited/.

3. "Statement by Scientists, Lawyers, and Public Policy Activists on Why We Need a Global Moratorium on the Creation of Potential Pandemic Pathogens (PPPs) Through Gain-of-Function Experiments," https://www.surveymonkey.com/r/XPJL2R9.

4. L. Kuo, G. J. Godeke, M. J. Raamsman, P. S. Masters, and P. J. Rottier, "Retargeting of Coronavirus by Substitution of the Spike Glycoprotein Ectodomain: Crossing the Host Cell Species Barrier," *Journal of Virology* 74, no. 3 (February 2000): 1393–406, https://doi .org/10.1128/JVI.74.3.1393-1406.2000.

5. Suryanarayanan, "Reading List: What Are the Origins of SARS-CoV-2?"

6. Carrey Gillam, "Validity of Key Studies on Origin of Coronavirus in Doubt; Science Journals Investigating," US Right to Know, November 9, 2020, https://www.organicconsumers.org /blog/validity-key-studies-origin-covid-in-doubt-science-journals-investigating.

7. Relman, "Opinion: To Stop the Next Pandemic."

8. Andrew Nikiforuk, "How China's Fails, Lies and Secrecy Ignited a Pandemic Explosion," *Tyee*, April 2, 2020, https://thetyee.ca/Analysis/2020/04/02/China-Secrecy-Pandemic/; Jeremy Page, Wenxin Fan, and Natasha Khan, "How It All Started: China's Early Coronavirus Missteps," *Wall Street Journal*, March 6, 2020, https://www.wsj.com/articles /how-it-all-started-chinas-early-coronavirus-missteps-11583508932; Steven Lee Myers, "China Created a Fail-Safe System to Track Contagions. It Failed," *New York Times*, March 29, 2020, https://www.nytimes.com/2020/03/29/world/asia/coronavirus-china.html.

9. *Preventing a Biological Arms Race*, ed. Susan Wright.

10. Lynn Kotz, "Human Error in High Biocontainment Labs: A Likely Pandemic Threat," *Bulletin of the Atomic Scientists*, February 25, 2019, https://thebulletin.org/2019/02/human -error-in-high-biocontainment-labs-a-likely-pandemic-threat/; Botao Xiaou, "The Possible Origins of the 2019-nCoV Coronavirus" (PDF), https://img-prod.tgcom24.mediaset.it /images/2020/02/16/114720192-5eb8307f-017c-4075-a697-348628da0204.pdf; Wang Keju, "Brucellosis Confirmed in 65 People from Lanzhou Veterinary Institute," China-Daily.com, updated December 16, 2019, accessed December 15, 2020, https://global .chinadaily.com.cn/a/201912/06/WS5deb4fe7a310cf3e3557c92a.html.

11. Committee on Anticipating Biosecurity Challenges of the Global Expansion of High-Containment Biological Laboratories et al., *Biosecurity Challenges of the Global Expansion of High-Containment Biological Laboratories: Summary of a Workshop* (Washington, DC: National Academies Press, 2011), chapter 1, https://doi.org/10.17226/13315; Ian Sample, "Revealed: 100 Safety Breaches at UK Labs Handling Potentially Deadly Disease," *Guardian*, December 4, 2014, https://www.theguardian.com/science/2014/dec/04/-sp-100

-safety-breaches-uk-labs-potentially-deadly-diseases; Natalie Vestin, "Federal Report Discloses Incidents in High-Containment Labs," CIDRAP, July 1, 2016, http://www .cidrap.umn.edu/news-perspective/2016/07/federal-report-discloses-incidents-high -containment-labs; Sharon Begley and Julie Steenhuysen, "How Secure Are Labs Handling World's Deadliest Pathogens?," Reuters, February 15, 2012, https://www.reuters.com /article/us-health-biosecurity-idUSTRE81E0R420120215; Lisa Schnirring, "CDC Monitoring More Staff After Anthrax Lab Breach," CIDRAP, June 20, 2014, http://www.cidrap .umn.edu/news-perspective/2014/06/cdc-monitoring-more-staff-after-anthrax-lab-breach; Christina Lin, "Biosecurity in Question at US Germ Labs," *Asia Times*, April 6, 2020, https://asiatimes.com/2020/04/biosecurity-in-question-at-us-germ-labs/; Jocelyn Kaiser, "Accidents Spur a Closer Look at Risks at Biodefense Labs," *Science* 317 (September 28, 2007): 1852–54, https://science.sciencemag.org/content/317/5846/1852?ck=nck.

12. Allison Young, "Newly Disclosed CDC Biolab Failures 'Like a Screenplay for a Disaster Movie,'" *USA Today*, June 2, 2016, https://www.usatoday.com/story/news/2016/06/02 /newly-disclosed-cdc-lab-incidents-fuel-concerns-safety-transparency/84978860/; Arthur Trapotsis, "Do You Know the Difference in Laboratory Biosafety Levels 1, 2, 3 & 4?," Consolidated Sterilizer Systems, updated March 31, 2020, accessed December 15, 2020, https://consteril.com/biosafety-levels-difference/.

13. Henry Fountain, "Six Vials of Smallpox Discovered in Laboratory Near Washington," *New York Times*, July 9, 2014, https://www.nytimes.com/2014/07/09/science/six-vials-of -smallpox-discovered-in-laboratory-near-washington.html.

14. Elisabeth Eaves, "Hot Zone in the Heartland," *Bulletin of the Atomic Scientists*, accessed December 15, 2020, https://thebulletin.org/2020/03/hot-zone-in-the-heartland/; Matt Field, "Experts Know the New Coronavirus Is Not a Bioweapon. They Disagree on Whether It Could Have Leaked from a Research Lab," *Bulletin of the Atomic Scientists*, March 30, 2020, https://thebulletin.org/2020/03/experts-know-the-new-coronavirus-is -not-a-bioweapon-they-disagree-on-whether-it-could-have-leaked-from-a-research-lab/.

15. US Government Accountability Office, "High-Containment Laboratories Improved Oversight of Dangerous Pathogens Needed to Mitigate Risk," GAO.gov, August 2016 (PDF), https://www.gao.gov/assets/680/679392.pdf.

16. Joe Lauria, "Worries About a Galveston Bio-Lab," Consortium News, August 30, 2017, https://consortiumnews.com/2017/08/30/worries-about-a-galveston-bio-lab/.

17. Denise Grady, "Deadly Germ Research Is Shut Down at Army Lab Over Safety Concerns," *New York Times*, August 5, 2020, https://www.nytimes.com/2019/08/05/health /germs-fort-detrick-biohazard.html.

18. David Cyranoski, "Inside the Chinese Lab Poised to Study World's Most Dangerous Pathogens," *Nature*, February 22, 2017, https://www.nature.com/news/inside-the-chinese -lab-poised-to-study-world-s-most-dangerous-pathogens-1.21487.

19. Grace Panetta, "US Officials Were Reportedly Concerned That Safety Breaches at Wuhan Lab Studying Coronavirus in Bats Could Cause a Pandemic," *Business Insider*, April 14, 2020, https://www.businessinsider.com/us-officials-raised-alarms-about-safety-issues-in -wuhan-lab-report-2020-4.

20. Fred Guterl, "Dr. Fauci Backed Controversial Wuhan Lab with US Dollars for Risky Coronavirus Research," *Newsweek*, April 28, 2020, https://www.newsweek.com/dr-fauci -backed-controversial-wuhan-lab-millions-us-dollars-risky-coronavirus-research-1500741.

21. NIH Project Information, "Understanding the Risk of Bat Coronavirus Emergence," Research Portfolio Online Reporting Tools, accessed December 1, 2020, https://projectreporter.nih.gov/project_info_description.cfm?aid=8674931&icde=49750546.

22. Vineet D. Menachery et al., "A SARS-Like Cluster of Circulating Bat Coronaviruses Shows Potential for Human Emergence," *Nature Medicine* 21 (2015): 1508–13, https://www.nature.com/articles/nm.3985.

23. Husseini, "Did This Virus Come from a Lab?"

24. Kevin Baker, "Did America Use Bioweapons in Korea? Nicholson Baker Tried to Find Out," *New York Times*, July 21, 2020, https://www.nytimes.com/2020/07/21/books/review/baseless-nicholson-baker.html.

25. Philip Sherwell, "Chinese Scientists Destroyed Proof of Virus in December," *Sunday Times*, March 1, 2020, https://www.thetimes.co.uk/article/chinese-scientists-destroyed-proof-of-virus-in-december-rz055qjnj.

26. Alexis Baden-Mayer, "Shi Zhengli: Weaponizing Coronaviruses, with Pentagon Funding, at a Chinese Military Lab," Organic Consumers Association, September 24, 2020, https://www.organicconsumers.org/blog/shi-zhengli-weaponizing-coronaviruses-pentagon-funding-chinese-military-lab.

27. Moreno Colaiacovo, "Fearsome Viruses and Where to Find Them," Medium.com, November 15, 2020, https://mygenomix.medium.com/fearsome-viruses-and-where-to-find-them-4e6b0ac6e602.

28. Alina Chan, Twitter thread, October 25, 2020, https://threadreaderapp.com/thread/1320344055230963712.html.

29. Jonathan Bucks, "New Cover-Up Fears as Chinese Officials Delete Critical Data About the Wuhan Lab with Details of 300 Studies Vanishing—Including All Those Carried Out by Virologist Dubbed Batwoman," *Daily Mail*, January 9, 2021, https://www.dailymail.co.uk/news/article-9129681/amp/New-cover-fears-Chinese-officials-delete-critical-data-Wuhan-lab.html.

30. Charles Calisher, "Statement in Support of the Scientists, Public Health Professionals, and Medical Professionals of China Combatting COVID-19," *Lancet* 395 (March 7, 2020): E42–E43, https://www.thelancet.com/journals/lancet/article/PIIS0140-6736(20)30418-9/fulltext.

31. Sainath Suryanarayanan, "EcoHealth Alliance Orchestrated Key Scientists' Statement on 'Natural Origin' of SARS-CoV-2," US Right to Know, November 18, 2020, https://usrtk.org/biohazards-blog/ecohealth-alliance-orchestrated-key-scientists-statement-on-natural-origin-of-sars-cov-2/; Jonathan Matthews, "EcoHealth Alliance Orchestrated Key Scientists' Statement on 'Natural Origin' of SARS-CoV-2," GM Watch, November 19, 2020, https://www.gmwatch.org/en/news/latest-news/19600.

32. Peter Daszak of EcoHealth Allience, email, February 6, 2020, https://usrtk.org/wp-content/uploads/2020/11/The_Lancet_Emails_Daszak-2.6.20.pdf.

33. Rita Colwell, email to Peter Daszak, February 8, 2020, https://usrtk.org/wp-content/uploads/2020/11/The_Lancet_Emails_Daszak-2.8.20.pdf.

34. Peter Daszak, "Members of the Lancet COVID Commission Task Force on the Origins of SARS-CoV-2 Named," EcoHealth Alliance, November 23, 2020, https://www.ecohealthalliance.org/2020/11/members-of-the-lancet-covid-commission-task-force-on-the-origins-of-sars-cov-2-named.

35. World Health Organization, "Origins of the SARS-CoV-2 Virus," updated January 18, 2021, https://www.who.int/health-topics/coronavirus/who-recommendations-to-reduce-risk-of -transmission-of-emerging-pathogens-from-animals-to-humans-in-live-animal-markets.

36. Betty L. Louie, Yufeng (Ethen) Ma, and Martha Wang, "China Proposes to Tighten Biosecurity Law and Its Potential Impact on Foreign Pharmaceutical and Biotech Companies Operating in China," Orrick.com, July 10, 2020, https://www.orrick.com/en/Insights/2020/07/China -Proposes-to-Tighten-Biosecurity-Law-and-its-Potential-Impact-on-Foreign-Companies.

37. Chaolin Huang et al., "Clinical Features of Patients Infected with 2019 Novel Coronavirus in Wuhan, China," *Lancet* 395, no. 10223 (February 15, 2020): 497–506, https://dpoi.org /10.1016/S0140-6736(20)30183-5.

38. Frank Chen, "Coronavirus 'Lab Leakage' Rumors Spreading," *Asia Times*, February 17, 2020, http://asiatimes.com/2020/02/coronavirus-lab-leakage-rumors-spreading.

39. Botao Xiaou, "The Possible Origins of the 2019-nCoV Coronavirus."

40. Botao Xiaou, "The Possible Origins of the 2019-nCoV Coronavirus."

41. Baden-Mayer, "Shi Zhengli."

42. Guterl, "Dr. Fauci Backed Controversial Wuhan Lab"; Editors, "Gain-of-Function Hall of Shame," Organic Consumers Association, ongoing, accessed December 15, 2020, https:// www.organicconsumers.org/news/gain-of-function-hall-of-shame.

43. Robert F. Kennedy, Jr., Instagram post, April 14, 2020, https://www.instagram.com/p /B--PXQKHxhs/.

44. Klotz, "Human Error in Bio-Containment Labs."

45. Dennis Normille, "Lab Accidents Prompt Calls for New Containment Program," *Science* 304, no. 5675 (May 28, 2004): 1223–25, https://doi.org/10.1126/science.304.5675.1223a.

46. Josh Rogin, "Commentary: State Department Cables Warned of Safety Issues at Wuhan Lab Studying Bat Coronaviruses," *Washington Post*, republished in *Bend Bulletin*, April 14, 2020, https://www.bendbulletin.com/opinion/commentary-state-department-cables -warned-of-safety-issues-at-wuhan-lab-studying-bat-coronaviruses/article_8ffbfcf2-7e79 -11ea-b101-cbcc5394b481.html.

47. Ian Birrell, "Beijing Now Admits That Coronavirus DIDN'T Start in Wuhan's Market . . . So Where DID It Come From, Asks IAN BIRRELL," *Daily Mail*, May 30, 2020, https:// www.dailymail.co.uk/news/article-8373007/Beijing-admits-coronavirus-DIDNT-start -Wuhans-market-DID-come-from.html?ITO=applenews.

48. Chaolin Huang et al., "Clinical Features of Patients Infected with 2019 Novel Coronavirus."

49. Roger Frutos et al., "COVID-19: Time to Exonerate the Pangolin from the Transmission of SARS-CoV-2 to Humans," *Infections, Genetics and Evolution* 84 (October 2020): 104493, https://doi.org/10.1016/j.meegid.2020.104493.

50. Colaiacovo, "Fearsome Viruses and Where to Find Them."

51. Shing Hei Zhan, Benjamin E. Deverman, and Yujia Alina Chan, "SARS-CoV-2 Is Well Adapted for Humans. What Does This Mean for Re-Emergence?," bioRxiv preprint, May 2, 2020, https://doi.org/10.1101/2020.05.01.073262.

52. Sakshi Piplani et al., "In Silico Comparison of Spike Protein-ACE2 Binding Affinities Across Species; Significance for the Possible Origin of the SARS-CoV-2 Virus," arXiv:2005.06199 [q-bio.BM] preprint, May 2020, https://arxiv.org/abs/2005.06199.

53. Jon Cohen, "Wuhan Coronavirus Hunter Shi Zhengli Speaks Out," *Science*, July 31, 2020, https://science.sciencemag.org/content/369/6503/487.full; Dr. Shi Zhengli, "Reply to

Science Magazine," accessed December 21, 2020, https://www.sciencemag.org/sites/default /files/Shi%20Zhengli%20Q&A.pdf.

54. Jocelyn Kaiser, "NIH Lifts 3-Year Ban on Funding Risky Virus Studies," *Science*, December 19, 2017, https://www.sciencemag.org/news/2017/12/nih-lifts-3-year-ban -funding-risky-virus-studies.

55. Editors, "Gain-of-Function Hall of Shame," Organic Consumers Association, ongoing, accessed December 15, 2020.

56. André Leu, "COVID 19: The Spike and the Furin Cleavage," Organic Consumers Association, June 3, 2020, https://www.organicconsumers.org/blog/covid-19-spike-and-furin-cleavage.

57. Max Roser, "The Spanish Flu (1918–20): The Global Impact of the Largest Influenza Pandemic in History," Our World in Data, March 4, 2020, https://ourworldindata.org /spanish-flu-largest-influenza-pandemic-in-history.

58. Alice Daniel, "Report to US Senator Durkin," GAO.gov, January 14, 1981 (PDF), https:// www.gao.gov/assets/140/132011.pdf.

59. Content Team, "The 1976 Swine Flu Vaccine Debacle—A Cautionary Tale for 2020," *Sault Online*, May 4, 2020, https://saultonline.com/2020/05/the-1976-swine-flu-vaccine -debacle-a-cautionary-tale-for-2020.

60. "Swine Flu 1976 Vaccine Warning, Part 1 of 2," *60 Minutes*, July 15, 2009, https://www .youtube.com/watch?v=VxeKY-TLmFk.

61. Geoff Earle, "'2 Million Dead'—Feds Make Chilling Forecast If Bird-Flu Pandemic Hits US," *New York Post*, May 4, 2006, https://nypost.com/2006/05/04/2-million-dead-feds -make-chilling-forecast-if-bird-flu-pandemic-hits-u-s/.

62. World Health Organization, "Safety of Pandemic Vaccines: Pandemic (H1N1) 2009 Briefing Note 6," August 6, 2009, https://www.who.int/csr/disease/swineflu/notes/h1n1 _safety_vaccines_20090805/en/.

63. Delece Smith-Barrow, "CDC's Advice to Parents: Swine Flu Shots for All," *Washington Post*, August 25, 2009, https://www.washingtonpost.com/wp-dyn/content/article/2009 /08/24/AR2009082402327.html.

64. *Eurosurveillance* Editorial Team, "Swedish Medical Products Agency Publishes Report from a Cast Inventory Study on Pandemrix Vaccination and Development of Narcolepsy with Cataplexy," *Eurosurveillance* 16, no. 26 (June 30, 2011), https://www.eurosurveillance .org/content/10.2807/ese.16.26.19904-en.

65. European Centre for Disease Prevention and Control, "Narcolepsy in Association with Pandemic Influenza Vaccination—a Multi-Country European Epidemiological Investiga- tion," September 20, 2012, https://www.ecdc.europa.eu/en/publications-data/narcolepsy -association-pandemic-influenza-vaccination-multi-country-european; Lisa Schnirring, "Study Funds Post-H1N1-Vaccination Rise in Narcolepsy in 3 Nations," CIDRAP, January 30, 2013, http://www.cidrap.umn.edu/news-perspective/2013/01/study-finds-post-h1n1 -vaccination-rise-narcolepsy-3-nations.

66. Pär Hallberg et al., "Pandemrix-Induced Narcolepsy Is Associated with Genes Related to Immunity and Neuronal Survival," *EBioMedicine* 40 (February 2019): 595–604, https:// www.ncbi.nlm.nih.gov/pmc/articles/PMC6413474/.

67. S. M. Zimmer and D. S. Burke, "Historical Perspective—Emergence of Influenza A (H1N1) Viruses," *New England Journal of Medicine* 361 (2009): 279–85, https://doi.org /10.1056/NEJMra0904322.

68. C. Scholtissek, V. von Hoyningen, and R. Rott, "Genetic Relatedness Between the New 1977 Epidemic Strains (H1N1) of Influenza and Human Influenza Strains Isolated Between 1947 and 1957 (H1N1)," *Virology* 89 (1978): 613–17.

69. R. G. Webster, W. J. Bean, O. T. Gorman, T. M. Chambers, and Y. Kawaoka, "Evolution and Ecology of Influenza A Viruses," *Microbiological Reviews* 56 (1992): 152–79; A. P. Kendal et al., "Antigenic Similarity of Influenza A (H1N1) Viruses from Epidemics in 1977–1978 to 'Scandinavian' Strains Isolated in Epidemics of 1950–1951," *Virology* 89 (197): 632–36.

70. Gerard Gallagher, "Fauci: '"No Doubt' Trump Will Face Surprise Infectious Disease Outbreak," *Infectious Disease News*, January 11, 2017, https://www.healio.com/news/infectious -disease/20170111/fauci-no-doubt-trump-will-face-surprise-infectious-disease-outbreak.

71. Holland, "What Can We Learn from a Pandemic Tabletop Exercise?"; Editors, "Press Pause," Organic Consumers Association, accessed December 16, 2020, https://www .organicconsumers.org/newsletter/we-need-regenerative-hero/press-pause; Joseph Mercola, "Hope Despite Censorship," Mercola.com, November 6, 2020, https://articles.mercola .com/sites/articles/archive/2020/11/06/hope-despite-censorship.aspx.

72. Colaiacovo, "Fearsome Viruses and Where to Find Them."

73. Klotz, "The Biological Weapons Convention Protocol Should Be Revisited."

74. William Gittins, "Bill Gates Predicts When the Next Pandemic Will Arrive," *AS*, December 15, 2020, https://en.as.com/en/2020/11/24/latest_news/1606228590_532670.html; Christopher Rosen, "Bill Gates Gives Stephen Colbert a Realistic Coronavirus Vaccine Timeline," *Vanity Fair*, April 24, 2020, https://www.vanityfair.com/hollywood/2020/04 /bill-gates-stephen-colbert-coronavirus-vaccine.

Chapter Three: Event 201 and the Great Reset

1. Alice Miranda Ollstein, "Trump Halts Funding to World Health Organization," *Politico*, April 14, 2020, https://www.politico.com/news/2020/04/14/trump-world-health -organization-funding-186786.

2. Josephine Moulds, "How Is the World Health Organization Funded?," World Economic Forum, April 15, 2020, https://www.weforum.org/agenda/2020/04/who-funds-world -health-organization-un-coronavirus-pandemic-covid-trump.

3. "World Leaders Commit to GAVI's Vision to Protect the Next Generation with Vaccines," Gavi, January 23, 2020, https://www.gavi.org/news/media-room/world-leaders-commit -gavis-vision-protect-next-generation-vaccines.

4. Mercola, "The Global Takeover Is Underway."

5. Steven Guinness, "Sustainable Chaos: When Globalists Call for a 'Great Reset,'" Technocracy.news, June 25, 2020, https://www.technocracy.news/sustainable-chaos-when-globalists -call-for-a-great-reset/.

6. Matt Hancock, Speech to the All-Part Parliamentary Group, "The Fourth Industrial Revolution," Gov.UK, October 16, 2017, https://www.gov.uk/government/speeches/the -4th-industrial-revolution.

7. Department of Global Communications, "Climate Change and COVID-19: UN Urges Nations to 'Recover Better,'" UN.org, April 22, 2020, https://www.un.org/en/un -coronavirus-communications-team/un-urges-countries-%E2%80%98build-back-better %E2%80%99; Mark Tovey, "Why Biden and Boris Are Both Using 'Build Back Better,'"

Intellectual Takeout, October 12, 2020, https://www.intellectualtakeout.org/why-biden
-and-boris-are-both-using--build-back-better-/.

8. Ida Auken, "Welcome to 2030: I Own Nothing, Have No Privacy and Life Has Never Been
Better," *Forbes*, November 10, 2016, https://www.forbes.com/sites/worldeconomicforum/2016
/11/10/shopping-i-cant-really-remember-what-that-is-or-how-differently-well-live-in-2030/.

9. "WO/2020/060606—Cryptocurrency System Using Body Activity Data," WIPO, March 26,
2020, https://patentscope.wipo.int/search/en/detail.jsf?docId=WO2020060606.

10. Tim Schwab, "Bill Gates's Charity Paradox," *Nation*, March 17, 2020, https://www
.thenation.com/article/society/bill-gates-foundation-philanthropy/.

11. Steerpike, "Six Questions That Neil Ferguson Should Be Asked," *Spectator*, April 16, 2020,
https://www.spectator.co.uk/article/six-questions-that-neil-ferguson-should-be-asked;
Saifedean Ammous, Twitter thread, May 3, 2020, https://twitter.com/saifedean/status
/1257101783408807938?s=21.

12. Steerpike, "Six Questions That Neil Ferguson Should Be Asked."

13. David Adam, "Special Report: The Simulations Driving the World's Response to COVID-19,"
Nature, April 2, 2020, https://www.nature.com/articles/d41586-020-01003-6.

14. Schwab, "Bill Gates's Charity Paradox."

15. Michael A. Rodriguez, MD, and Robert García, JD, "First, Do No Harm: The US Sexually
Transmitted Disease Experiments in Guatemala," *American Journal of Public Health* 103, no.
12 (December 2013): 2122–26, https://dx.doi.org/10.2105%2FAJPH.2013.301520.

16. Meridian 361 International Law Group, PLLC, "Rockefeller, Johns Hopkins Behind
Horrific Human Syphilis Experiments, Allege Guatemalan Victims In Lawsuit," Cision,
PR Newswire, April 1, 2015, https://www.prnewswire.com/news-releases/rockefeller
-johns-hopkins-behind-horrific-human-syphilis-experiments-allege-guatemalan-victims
-in-lawsuit-300059537.html.

17. Chuck Ross, "World Health Organization Hired PR Firm to Identify Celebrity 'Influencers' to Amplify Virus Messaging," *Daily Caller*, July 17, 2020, https://dailycaller.com/2020
/07/17/world-health-organization-coronavirus-celebrity-influencers.

18. US Department of Justice, "Exhibit A Registration Statement," Foreign Agents Registration Act, July 14, 2020, https://efile.fara.gov/docs/3301-Exhibit-AB-20200714-38.pdf.

19. Verified, accessed December 22, 2020, https://content.shareverified.com/en; Dr. Joseph
Mercola, "The PR Firm Behind WHO's Celeb Endorsements," Mercola.com, August 15,
2020, https://articles.mercola.com/sites/articles/archive/2020/08/15/world-health
-organization-endorsements.aspx.

20. Publicis Groupe, World Economic Forum, https://www.weforum.org/organizations
/publicis-groupe-sa.

21. Publicis Groupe, "Publicis Groupe Acquires Remaining Capital of Leo Burnett/W&K
Beijing Advertising Co., Ltd.," Publicis Groupe, April 29, 2010, https://www.publicisgroupe
.com/sites/default/files/press-release/20100429_10-04-29_LeoB_and_W%26K_ENG_DEF
.pdf; "Our Investors," NewsGuard, https://www.newsguardtech.com/about/our-investors;
Tom Burt, "Defending Against Disinformation in Partnership with NewsGuard," *Microsoft
on the Issues* (blog), August 23, 2018, https://blogs.microsoft.com/on-the-issues/2018/08/23
/defending-against-disinformation-in-partnership-with-newsguard.

22. Rachel Blevins, "Ron Paul: Police State Was Planned, 9/11 Just 'Provided an Opportunity' to
Implement It," Free Thought Project, September 11, 2017, https://thefreethoughtproject.com

/ron-paul-patriot-act-911; Judge Andrew P. Napolitano, "The Patriot Act Must Go: It Assaults Our Freedoms, Doesn't Keep Us Safe," Fox News, last updated May 29, 2015, https://www .foxnews.com/opinion/the-patriot-act-must-go-it-assaults-our-freedoms-doesnt-keep-us-safe.

23. "Surveillance Under the Patriot Act," ACLU, https://www.aclu.org/issues/national-security /privacy-and-surveillance/surveillance-under-patriot-act.

24. Ariel Zilber, "How Bill Gates Warned in 2015 TED Talk That the Next Big Threat to Humanity Was a 'Highly Infectious Virus' That 'We Are Not Ready' For," DailyMail.com, March 19, 2020, https://www.dailymail.co.uk/news/article-8132107/Bill-Gates-warned -2015-TED-Talk-big-threat-humanity-coronavirus-like-pandemic.html.

25. James Corbett, "Who Is Bill Gates?" The Corbett Report, May 1, 2020, https://www .corbettreport.com/gates/.

26. Justin Fitzgerald, "IMF Calls for Credit Score to Be Tied to Internet Search History," Reality Circuit, December 23, 2020, https://realitycircuit.com/2020/12/23/imf-calls -for-credit-score-to-be-tied-to-internet-search-history/.

27. Ellen Sheng, "Facebook, Google Discuss Sharing Smartphone Data with Government to Fight Coronavirus, but There Are Risks," CNBC, March 19, 2020, https://www.cnbc.com /2020/03/19/facebook-google-could-share-smartphone-data-to-fight-coronavirus.html.

28. Aaron Holmes, "Facebook Built a Tool Last Year to Map the Spread of Diseases. Now It's Being Used to Combat Coronavirus. Here's How It Works," Business Insider, March 18, 2020, https://www.businessinsider.com/see-how-facebooks-disease-prevention-maps -could-fight-coronavirus-2020-3.

29. "Disease Prevention Maps," Facebook Data for Good, https://dataforgood.fb.com/tools /disease-prevention-maps.

30. Kevin Granville, "Facebook and Cambridge Analytica: What You Need to Know as Fallout Widens," New York Times, March 19, 2018, https://www.nytimes.com/2018/03/19 /technology/facebook-cambridge-analytica-explained.html.

31. Matt Perez, "Bill Gates Calls for National Tracking System for Coronavirus During Reddit AMA," Forbes, March 18, 2020, https://www.forbes.com/sites/mattperez/2020/03/18 /bill-gates-calls-for-national-tracking-system-for-coronavirus-during-reddit-ama /?sh=737b58726a72.

32. Bill Gates, "31 Questions and Answers About COVID-19," GatesNotes (blog), March 19, 2020, https://www.gatesnotes.com/Health/A-coronavirus-AMA.

33. World Health Organization and World Economic Forum, "Preventing Noncommunicable Diseases in the Workplace Through Diet and Physical Activity," World Health Organization, 2008, https://www.who.int/dietphysicalactivity/WHOWEF_report_JAN2008_FINAL.pdf.

34. "The Great Reset Launch Session," World Economic Forum, June 3, 2020 (video), https:// www.youtube.com/watch?v=pfVdMWzKwjc&t=1s.

35. Klaus Schwab and Theirry Malleret, The Great Reset (Geneva, Switzerland: World Economic Forum, 2020), 173.

36. Declan McCullagh, "Joe Biden's Pro-RIAA, Pro-FBI Tech Voting Record," CNET, August 24, 2008, https://www.cnet.com/news/joe-bidens-pro-riaa-pro-fbi-tech-voting-record.

37. Department of Global Communications, "Climate Change and COVID-19."

38. Department of Global Communications, "Climate Change and COVID-19."

39. "The Campaign for a Coronavirus Recovery Plan That Builds Back Better," Build Back Better, accessed December 20, 2020, https://www.buildbackbetteruk.org.

40. Anna North, "New Zealand Prime Minister Jacinda Ardern Wins Historic Reelection," *Vox*, October 17, 2020, https://www.vox.com/2020/10/17/21520584/jacinda-ardern -new-zealand-prime-minister-reelection-covid-19.

41. Paul Wong and Jesse Leigh Maniff, "Comparing Means of Payment: What Role for a Central Bank Digital Currency?," FEDS Notes, Federal Reserve, August 13, 2020, https:// www.federalreserve.gov/econres/notes/feds-notes/comparing-means-of-payment-what -role-for-a-central-bank-digital-currency-20200813.htm.

Chapter Four: COVID-19 Strikes the Most Vulnerable

1. "Provisional Death Counts for Coronavirus Disease 2019," Centers for Disease Control and Prevention, updated December 9, 2020, https://www.cdc.gov/nchs/nvss/vsrr/covid _weekly/index.htm.

2. Youyou Zhou and Gary Stix, "COVID-19 Is Now the Third Leading Cause of Death in the US," *Scientific American*, October 8, 2020, https://www.scientificamerican.com/article /covid-19-is-now-the-third-leading-cause-of-death-in-the-u-s1; Jackie Salo, "COVID-19 Is Third Leading Cause of Death in the United States," *New York Post*, August 18, 2020, https://nypost.com/2020/08/18/covid-19-is-third-leading-cause-of-death-in-the -united-states.

3. *Johns Hopkins News-Letter* (@JHUNewsLetter), "Though making clear the need for further research, the article was being used to support false and dangerous inaccuracies about the impact of the pandemic . . . ," Twitter, November 26, 2020, https://twitter.com /JHUNewsLetter/status/1332100155986882562.

4. Yanni Gu, "A Closer Look at US Deaths Due to COVID-19," *Johns Hopkins News-Letter* web archive, November 22, 2020, https://web.archive.org/web/20201126163323 /https:/www.jhunewsletter.com/article/2020/11/a-closer-look-at-u-s-deaths-due-to -covid-19.

5. Ethan Yang, "New Study Highlights Alleged Accounting Error Regarding Covid Deaths," American Institute for Economic Research, November 26, 2020, https://www.aier.org /article/new-study-highlights-serious-accounting-error-regarding-covid-deaths.

6. Stephen Schwartz, PhD, "Guidance for Certifying COVID-19 Deaths," Centers for Disease Control and Prevention, National Center for Health Statistics, Division of Vital Statistics, March 4, 2020, https://www.cdc.gov/nchs/data/nvss/coronavirus/alert-1-guidance-for -certifying-COVID-19-deaths.pdf.

7. National Vital Statistics System, "Guidance for Certifying Deaths Due to Coronavirus Disease 2019 (COVID-19)," Vital Statistics Reporting Guidance, report number 3, April 2020, https://www.cdc.gov/nchs/data/nvss/vsrg/vsrg03-508.pdf.

8. National Center for Health Statistics, "COVID-19 Death Data and Resources," Centers for Disease Control and Prevention, last reviewed November 25, 2020, accessed December 8, 2020, https://www.cdc.gov/nchs/nvss/covid-19.htm.

9. Tanya Lewis, "Eight Persistent COVID-19 Myths and Why People Believe Them," *Scientific American*, October 12, 2020, https://www.scientificamerican.com/article/eight-persistent -covid-19-myths-and-why-people-believe-them/.

10. Justin Blackburn, PhD, et al., "Infection Fatality Ratios for COVID-19 Among Noninsti-tutionalized Persons 12 and Older: Results of a Random-Sample Prevalence Study," *Annals of Internal Medicine* (2020), https://www.acpjournals.org/doi/10.7326/M20-5352.

11. Lee Merritt, MD, "SARS-CoV2 and the Rise of Medical Technocracy," August 16, 2020 (video), DDP 38th Annual Meeting, Las Vegas, Nevada, https://www.youtube.com/watch?v=sjYvitCeMPc&feature=emb_title.
12. Frank E. Lockwood and John Moritz, "Birx Says Country Weary of COVID-19, Recognizes Arkansas' Improvement During Visit," *El Dorado News-Times*, August 18, 2020, https://www.eldoradonews.com/news/2020/aug/18/birx-says-country-weary-covid-19-recognizes-arkans/?fbclid=IwAR07eHiJSLp6UPXd6dabokayamMiXV5aR4EOxROiEuUCf3_5ikKHMXLNGko.
13. Associated Press, "Navy ID's USS *Roosevelt* Sailor Killed by COVID-19," NBC San Diego online, April 16, 2020, https://www.nbcsandiego.com/news/local/military/navy-ids-uss-roosevelt-sailor-killed-by-covid-19/2307424; Nick Givas and Lucas Tomlinson, "USS *Theodore Roosevelt*'s Entire Crew Has Been Tested for Coronavirus; Over 800 Positive, Officials Say," Fox News online, April 23, 2020, https://www.foxnews.com/world/uss-theodore-roosevelt-entire-crew-tested-coronavirus; Ryan Pickrell, "Sweeping US Navy Testing Reveals Most Aircraft Carrier Sailors Infected with Coronavirus Had No Symptoms," *Business Insider*, April 17, 2020, https://www.businessinsider.com/testing-reveals-most-aircraft0-carrier-sailors-coronavirus-had-no-symptoms-2020-4.
14. Centers for Disease Control and Prevention, "Public Health Responses to COVID-19 Outbreaks on Cruise Ships—Worldwide, February–March 2020," *Morbidity and Mortality Weekly Report* 69, no. 12 (2020): 347–52, https://www.cdc.gov/mmwr/volumes/69/wr/mm6912e3.htm.
15. Ray Sipherd, "The Third-Leading Cause of Death in US Most Doctors Don't Want You to Know About," CNBC, February 22, 2018, https://www.cnbc.com/2018/02/22/medical-errors-third-leading-cause-of-death-in-america.html?__source=sharebar%7Ctwitter&par=sharebar; Martin A. Makary and Michael Daniel, "Medical Error—The Third Leading Cause of Death in the US," *BMJ* 353 (2016): i2139, https://doi.org/10.1136/bmj.i2139.
16. John T. James, PhD, "A New, Evidence-Based Estimate of Patient Harms Associated with Hospital Care," *Journal of Patient Safety* 9, no. 3 (2013): 122–28, https://journals.lww.com/journalpatientsafety/fulltext/2013/09000/a_new,_evidence_based_estimate_of_patient_harms.2.aspx.
17. Martin Gould, "EXCLUSIVE: 'It's a Horror Movie.' Nurse Working on Coronavirus Frontline in New York Claims the City Is 'Murdering' COVID-19 Patients by Putting Them on Ventilators and Causing Trauma to the Lungs," DailyMail.com, April 27, 2020, https://www.dailymail.co.uk/news/article-8262351/nurse-new-york-claims-city-killing-COVID-19-patients-putting-ventilators.html.
18. Dr. Joseph Mercola, "CDC Admits Hospital Incentives Drove Up COVID-19 Deaths," Mercola.com, August 20, 2020, https://articles.mercola.com/sites/articles/archive/2020/08/20/hospital-incentives-drove-up-covid-19-deaths.aspx.
19. Matthew Boyle, "Exclusive—Seema Verma: Cuomo, Other Democrat Governors' Coronavirus Nursing Home Policies Contradicted Federal Guidance," *Breitbart*, June 22, 2020, https://www.breitbart.com/politics/2020/06/22/exclusive-seema-verma-cuomo-other-democrat-governors-coronavirus-nursing-home-policies-contradicted-federal-guidance/?fbclid=IwAR1XoAVEI4TzcoKbeVjbJk92WTdO6Qf-kPKTUIvotc9aDpcjW1VfrITW_RU.
20. Gregg Girvan, "Nursing Homes & Assisted Living Facilities Account for 42% of COVID-19 Deaths," Foundation for Research on Equal Opportunity, May 7, 2020, https://freopp.org/the-covid-19-nursing-home-crisis-by-the-numbers-3a47433c3f70; Avik Roy, "The

Most Important Coronavirus Statistic: 42% of US Deaths Are from 0.6% of the Population," *Forbes* online, May 26, 2020, https://www.forbes.com/sites/theapothecary/2020/05/26/nursing-homes-assisted-living-facilities-0-6-of-the-u-s-population-43-of-u-s-covid-19-deaths/?sh=30a0049f74cd.

21. OECD Policy Responses to Coronavirus, "Workforce and Safety in Long-Term Care During the COVID-19 Pandemic," Organization for Economic Cooperation and Development, June 22, 2020, https://www.oecd.org/coronavirus/policy-responses/workforce-and-safety-in-long-term-care-during-the-covid-19-pandemic-43fc5d50.

22. Bernadette Hogan and Bruce Golding, "Nursing Homes Have 'No Right' to Reject Coronavirus Patients, Cuomo Says," *New York Post*, April 23, 2020, https://nypost.com/2020/04/23/nursing-homes-cant-reject-coronavirus-patients-cuomo-says.

23. Joaquin Sapien and Joe Sexton, "Fire Through Dry Grass: Andrew Cuomo Saw COVID-19's Threat to Nursing Homes. Then He Risked Adding to It," ProPublica, June 16, 2020, https://www.propublica.org/article/fire-through-dry-grass-andrew-cuomo-saw-covid-19-threat-to-nursing-homes-then-he-risked-adding-to-it.

24. Arjen M. Dondorp et al., "Respiratory Support in COVID-19 Patients, with a Focus on Resource-Limited Settings," *American Journal of Tropical Medicine and Hygiene* 102, no. 6 (June 3, 2020): 1191–97, https://doi.org/10.4269/ajtmh.20-0283.

25. Giacomo Grasselli, MD, et al., "Baseline Characteristics and Outcomes of 1591 Patients Infected with SARS-CoV-2 Admitted to ICUs of the Lombardy Region, Italy," *JAMA* 323, no. 16 (2020): 1574–81, https://www.doi.org/10.1001/jama.2020.5394.

26. Safiya Richardson, MD, MPH, et al., "Presenting Characteristics, Comorbidities, and Outcomes Among 5700 Patients Hospitalized with COVID-19 in the New York City Area," *JAMA* 323, no. 20 (2020): 2052–59, https://www.doi.org/10.1001/jama.2020.6775.

27. Pavan K. Bhatraju, MD, et al., "Covid-19 in Critically Ill Patients in the Seattle Region—Case Series," *New England Journal of Medicine* 382 (2020): 2012–22, https://www.doi.org/10.1056/NEJMoa2004500.

28. "Sepsis," Centers for Disease Control and Prevention, accessed December 20 2020, https://www.cdc.gov/sepsis/clinicaltools/index.html?CDC_AA_refVal=https%3A%2F%2Fwww.cdc.gov%2Fsepsis%2Fdatareports%2Findex.html.

29. Vincent Liu, MD, MS, et al., "Hospital Deaths in Patients with Sepsis from 2 Independent Cohorts," *JAMA* 312, no. 1 (2014): 90–92, https://www.doi.org/10.1001/jama.2014.5804.

30. Richard Harris, "Stealth Disease Likely to Blame for 20 Percent of Global Deaths," NPR online, January 16, 2020, https://www.npr.org/sections/health-shots/2020/01/16/796758060/stealth-disease-likely-to-blame-for-20-of-global-deaths.

31. "Sepsis Alliance & Elara Caring Partner to Improve COVID-19 and Sepsis Outcomes in Home Healthcare Patients," Sepsis Alliance, July 1, 2020, https://www.sepsis.org/news/sepsis-alliance-elara-caring-partner-to-improve-covid-19-and-sepsis-outcomes-in-home-healthcare-patients.

32. Audrey Howard, "Covid-19 and Sepis Coding: New Guidelines," *Inside Angle from 3M Health Information Systems* (blog), April 2, 2020, https://www.3mhisinsideangle.com/blog-post/covid-19-and-sepsis-coding-new-guidelines.

33. Hui Li, MD, et al., "SARS-CoV-2 and Viral Sepsis: Observations and Hypotheses," *Lancet* 395, no. 10235 (May 9, 2020): 1517–20, https://www.doi.org/10.1016/S0140-6736(20)30920-X.

34. "Covid-19, Sepsis, and Cytokine Storms," Sepsis Alliance, May 20, 2020, https://www.sepsis.org/news/covid-19-sepsis-and-cytokine-storms.

35. "Report Sulle Caratteristiche dei Pazienti Deceduti Positivi a COVID-19 in Italia Il Presente Report è Basato sui Dati Aggiornati al 17 Marzo 2020," EpiCentro, March 17, 2020, https://www.epicentro.iss.it/coronavirus/bollettino/Report-COVID-2019_17_marzo-v2.pdf.

36. Centers for Disease Control and Prevention, "Hospitalization Rates and Characteristics of Patients Hospitalized with Laboratory-Confirmed Coronavirus Disease 2019—COVID-NET, 14 States, March 1–30, 2020," *Morbidity and Mortality Weekly Report* 69, no. 15 (April 17, 2020): 458–64, https://www.cdc.gov/mmwr/volumes/69/wr/mm6915e3.htm?s_cid=mm6915e3_w.

37. Richardson et al., "Presenting Characteristics, Comorbidities, and Outcomes."

38. Chao Gao et al., "Association of Hypertension and Antihypertensive Treatment with COVID-19 Mortality: A Retrospective Observational Study," *European Heart Journal* 41, no. 22 (June 7, 2020): 2058–66, https://doi.org/10.1093/eurheartj/ehaa433; Courtney Kueppers, "Study Shows High Blood Pressure Doubles Risk of Dying from COVID-19," *Atlanta Journal-Constitution*, June 5, 2020, https://www.ajc.com/lifestyles/study-shows-high-blood-pressure-doubles-risk-dying-from-covid/wUmZR3d52aBXJnEtilUnJK.

39. Deborah J. Nelson, "Blood-Pressure Drugs Are in the Crosshairs of COVID-19 Research," Reuters, April 23, 2020, https://www.reuters.com/article/us-health-conoravirus-blood-pressure-ins/blood-pressure-drugs-are-in-the-crosshairs-of-covid-19-research-idUSKCN2251GQ.

40. Deborah J. Nelson, "Blood-Pressure Drugs Are in the Crosshairs of COVID-19 Research," *Reuters*, April 23, 2020, https://www.reuters.com/article/us-health-conoravirus-blood-pressure-ins-idUSKCN2251GQ.

41. A. B. Docherty et al., "Features of 16,749 Hospitalised UK Patients with COVID-19 Using the ISARIC WHO Clinical Characterisation Protocol," medRxiv preprint, accessed December 20, 2020, https://doi.org/10.1101/2020.04.23.20076042.

42. "Diabetes Prevalence," Diabetes.co.uk, January 15, 2019, https://www.diabetes.co.uk/diabetes-prevalence.html.

43. Arjen M. Dondorp et al., "Respiratory Support in COVID-19 Patients, with a Focus on Resource-Limited Settings," *American Journal of Tropical Medicine and Hygiene* 102, no. 6 (June 3, 2020): 1191–97, https://doi.org/10.4269/ajtmh.20-0283.

44. Jamie Hartmann-Boyce, "The Type of Diabetes You Have Can Impact How You React to the Coronavirus," Scroll.in, June 7, 2020, https://scroll.in/article/963807/the-type-of-diabetes-you-have-can-impact-how-you-react-to-the-coronavirus.

45. Weina Guo et al., "Diabetes Is a Risk Factor for the Progression and Prognosis of COVID-19," *Diabetes/Metabolism Research and Reviews* 36, no. 7 (2020): e3319, https://doi.org/10.1002/dmrr.3319.

46. Matteo Rottoli et al., "How Important Is Obesity as a Risk Factor for Respiratory Failure, Intensive Care Admission and Death in Hospitalised COVID-19 Patients? Results from a Single Italian Centre," *European Journal of Endocrinology* 183, no. 4 (October 2020): 389–97, https://www.doi.org/10.1530/EJE-20-0541.

47. Alan Mozes, "Even Mild Obesity Raises Odds for Severe COVID-19," *US News & World Report*, July 23, 2020, https://www.usnews.com/news/health-news/articles/2020-07-23/even-mild-obesity-raises-odds-for-severe-covid-19.

48. Public Health England, "Excess Weight and COVID-19: Insights from New Evidence," July 2020, https://assets.publishing.service.gov.uk/government/uploads/system/uploads/attachment_data/file/903770/PHE_insight_Excess_weight_and_COVID-19.pdf.

49. Fumihiro Sanada et al., "Source of Chronic Inflammation in Aging," *Frontiers in Cardiovascular Medicine* 5 (2018): 12, https://www.doi.org/10.3389/fcvm.2018.00012.

50. "Coronavirus Disease 2019: Older Adults," Centers for Disease Control and Prevention, updated December 13, 2020, https://www.cdc.gov/coronavirus/2019-ncov/need-extra -precautions/older-adults.html.

51. Amber L. Mueller, "Why Does COVID-19 Disproportionately Affect Older People?," *Aging* 12, no. 10 (May 29, 2020): 9959–81, https://doi.org/10.18632/aging.103344.

52. Mueller, "Why Does COVID-19 Disproportionately Affect Older People?"

Chapter Five: Exploiting Fear to Lock Down Freedom

1. "COVID-19," Organic Consumers Association, accessed December 20, 2020, https://www .organicconsumers.org/campaigns/covid-19.

2. Centers for Disease Control and Prevention, "Covid-19 Planning Scenarios, Table 1, Scenario 5: Current Best Estimate," updated September 10, 2020, https://www.cdc.gov /coronavirus/2019-ncov/hcp/planning-scenarios.html.

3. Collins, "US Billionaire Wealth Surges Past $1 Trillion Since Beginning of Pandemic."

4. Chuck Collins, "US Billionaire Wealth Surges to $584 Billion, or 20 Percent, Since the Beginning of the Pandemic," Institute for Policy Studies, June 18, 2020, https://ips-dc.org /us-billionaire-wealth-584-billion-20-percent-pandemic.

5. Gillian Friedman, "Big-Box Retailers' Profits Surge as Pandemic Marches On," *New York Times*, August 19, 2020, https://www.nytimes.com/2020/08/19/business/coronavirus -walmart-target-home-depot.html.

6. Dr. Asoka Bandarage, "Pandemic, 'Great Reset' and Resistance," Inter Press Service (IPS), December 1, 2020, https://www.ipsnews.net/2020/12/pandemic-great-reset-resistance.

7. Dr. Elke Van Hoof, "Lockdown Is the World's Biggest Psychological Experiment—And We Will Pay the Price," World Economic Forum, April 9, 2020, https://www.weforum.org/agenda /2020/04/this-is-the-psychological-side-of-the-covid-19-pandemic-that-were-ignoring.

8. Madeleine Ngo, "Small Businesses Are Dying by the Thousands—and No One Is Tracking the Carnage," *Bloomberg*, August 11, 2020, https://www.bloomberg.com/news /articles/2020-08-11/small-firms-die-quietly-leaving-thousands-of-failures-uncounted.

9. Anjali Sundaram, "Yelp Data Shows 60% of Business Closures Due to the Coronavirus Pandemic Are Now Permanent," CNBC, September 16, 2020, https://www.cnbc.com /2020/09/16/yelp-data-shows-60percent-of-business-closures-due-to-the-coronavirus -pandemic-are-now-permanent.html.

10. Sundaram, "Yelp Data Shows 60% of Business Closures."

11. Pedro Nicolaci da Costa, "The Covid-19 Crisis Has Wiped Out Nearly Half of Black Small Business," *Forbes*, August 10, 2020, https://www.forbes.com/sites/pedrodacosta/2020/08/10 /the-covid-19-crisis-has-wiped-out-nearly-half-of-black-small-businesses/?sh=5c79efb43108.

12. Claire Kramer Mills, PhD, and Jessica Battisto, "Double Jeopardy: Covid-19's Concentrated Health and Wealth Effects in Black Communities," Federal Reserve Bank of New York, August 2020, https://www.newyorkfed.org/medialibrary/media/smallbusiness /DoubleJeopardy_COVID19andBlackOwnedBusinesses.

13. Bethan Staton and Judith Evans, "Three Million Go Hungry in U.K. Because of Lockdown," *Financial Times*, April 10, 2020, https://www.ft.com/content/e5061be6-2978-4c0b -aa68-f372a2526826.

14. "Covid Pushes Millions More Children Deeper into Poverty, New Study Finds," UN News, September 17, 2020 https://news.un.org/en/story/2020/09/1072602.

15. Fiona Harvey, "Coronavirus Pandemic 'Will Cause Famine of Biblical Proportions,'" *Guardian*, April 21, 2020, https://www.theguardian.com/global-development/2020/apr/21/coronavirus-pandemic-will-cause-famine-of-biblical-proportions.

16. Morganne Campbell, "Canadians Reporting Higher Levels of Anxiety, Depression amid the Pandemic," *Global News Canada*, October 10, 2020, https://globalnews.ca/news/7391217/world-mental-health-day-canada/.

17. American Psychological Association, "Stress in America 2020: A National Mental Health Crisis," October 2020, https://www.apa.org/news/press/releases/stress/2020/report-october; Cory Stieg. "More than 7 in 10 Gen-Zers Report Symptoms of Depression During Pandemic, Survey Finds," CNBC, October 21, 2020, https://www.cnbc.com/2020/10/21/survey-more-than-7-in-10-gen-zers-report-depression-during-pandemic.html.

18. Advocacy Resource Center, "Issue Brief: Reports of Increases in Opioid- and Other Drug-Related Overdose and Other Concerns During COVID Pandemic," American Medical Association, updated December 9, 2020, https://www.ama-assn.org/system/files/2020-12/issue-brief-increases-in-opioid-related-overdose.pdf.

19. Lauren M. Rossen, PhD, et al., "Excess Deaths Associated with COVID-19, by Age and Race and Ethnicity—United States, January 26–October 3, 2020," *Morbidity and Mortality Weekly Report* 69, no. 42 (October 23, 2020): 1522–27, https://www.cdc.gov/mmwr/volumes/69/wr/mm6942e2.htm?s_cid=mm6942e2_w; Amanda Prestigiacomo, "New CDC Numbers Show Lockdown's Deadly Toll on Young People," *Daily Wire*, October 22, 2020, https://www.dailywire.com/news/new-cdc-numbers-show-lockdowns-deadly-toll-on-young-people.

20. Jack Power, "Covid-19: Reports of Rape and Child Sex Abuse Rise Sharply During Pandemic," *Irish Times*, July 20, 2020, https://www.irishtimes.com/news/social-affairs/covid-19-reports-of-rape-and-child-sex-abuse-rise-sharply-during-pandemic-1.4308307.

21. Stacy Francis, "Op-ed: Uptick in Domestic Violence Amid Covid-19 Isolation," CNBC, October 30, 2020, https://www.cnbc.com/2020/10/30/uptick-in-domestic-violence-amid-covid-19-isolation.html.

22. "Domestic Abuse Killings Double and Calls to Helpline Surge by 50% During Coronavirus Lockdown," ITV.com, April 27, 2020, https://www.itv.com/news/2020-04-27/domestic-abuse-killings-double-and-calls-to-helpline-surge-by-50-during-coronavirus-lockdown.

23. Alan Mozes, "Study Finds Rise in Domestic Violence During COVID," WebMD, August 18, 2020, https://www.webmd.com/lung/news/20200818/radiology-study-suggests-horrifying-rise-in-domestic-violence-during-pandemic#1.

24. "UN Chief Calls for Domestic Violence 'Ceasefire' amid 'Horrifying Global Surge,'" UN News, April 6, 2020, https://news.un.org/en/story/2020/04/1061052.

25. Nicola McAlley, "Calls to Domestic Abuse Helpline Double During Lockdown," STV.tv, July 1, 2020, https://news.stv.tv/highlands-islands/calls-to-domestic-abuse-helpline-double-during-lockdown?top.

26. Jai Sidpra, "Rise in the Incidence of Abusive Head Trauma During the COVID-19 Pandemic," *Archives of Disease in Childhood*, July 2, 2020, https://www.doi.org/10.1136/archdischild-2020-319872.

27. Perry Stein, "In DC, Achievement Gap Widens, Early Literacy Progress Declines During Pandemic, Data Show," *Washington Post*, October 30, 2020, https://www.washingtonpost

.com/local/education/data-indicate-worsening-early-literacy-progress-and-widening
-achievement-gap-among-district-students/2020/10/30/bebe2914-1a25-11eb-82db
-60b15c874105_story.html.

28. "Lockdowns Could Have Long-Term Effects on Children's Health," *Economist*, July 19, 2020, https://www.economist.com/international/2020/07/19/lockdowns-could-have-long
-term-effects-on-childrens-health.

29. SBG San Antonio, "HOSPITAL: 37 Children Attempted Suicide in September, Highest Number in Five Years," CBS Austin, October 27, 2020, https://cbsaustin.com/news/local
/cook-childrens-hospital-admits-alarming-rate-of-suicide-attempts-in-children.

30. Selina Wang, Rebecca Wright, and Yoko Wakatsuki, "In Japan, More People Died from Suicide Last Month than from Covid in All of 2020. And Women Have Been Impacted Most," CNN, November 30, 2020, https://edition.cnn.com/2020/11/28/asia/japan-suicide
-women-covid-dst-intl-hnk/index.html.

31. Nate Doromal, "Covid Antifragility: Trusting Our Strength in Uncertain Times," *Evolution of Medicine*, December 3, 2020, https://goevomed.com/blogs/covid-antifragility
-trusting-our-strength-in-uncertain-times.

32. Nate Doromal, "Covid Antifragility: Trusting Our Strength in Uncertain Times," Organic Consumers Association, December 14, 2020, https://www.organicconsumers.org/news
/covid-antifragility-trusting-our-strength-uncertain-times.

33. "Great Barrington Declaration," Great Barrington Declaration, accessed December 15, 2020, https://gbdeclaration.org.

34. Desmond Sutton, MD, et al., "Correspondence: Universal Screening for SARS-CoV-2 in Women Admitted for Delivery," *New England Journal of Medicine* 382 (April 13, 2020): 2163–64, https://www.nejm.org/doi/full/10.1056/NEJMc2009316.

35. Travis P. Baggett et al., "COVID-19 Outbreak at a Large Homeless Shelter in Boston: Implication for Universal Testing," medRxiv preprint, April 15, 2020, https://doi.org/10
.1101/2020.04.12.20059618.

36. Shiyi Cao et al., "Post-Lockdown SARS-CoV-2 Nucleic Acid Screening in Nearly Ten Million Residents of Wuhan, China," *Nature Communications* 11, article number 5917 (November 20, 2020), https://www.nature.com/articles/s41467-020-19802-w.

37. "Provisional Death Counts for Coronavirus Disease 2019," Centers for Disease Control and Prevention, updated December 9, 2020, https://www.cdc.gov/nchs/nvss/vsrr/covid
_weekly/index.htm.

38. Merritt, "SARS-CoV-2 and the Rise of Medical Technocracy," approximately eight minutes in (Lie No. 1: Death Risk); D. G. Rancourt, "All-Cause Mortality During COVID-19: No Plague and a Likely Signature of Mass Homicide by Government Response," *Technical Report*, June 2020, https://www.doi.org/10.13140/RG.2.24350.77125; Yanni Gu, "A Closer Look at US Deaths Due to COVID-19," *Johns Hopkins News-Letter*, November 22, 2020 (archived), https://web.archive.org/web/20201126163323/https:
/www.jhunewsletter.com/article/2020/11/a-closer-look-at-u-s-deaths-due-to-covid-19.

39. "Fauci Says Schools Should Try to Stay Open," Mercola.com, December 27, 2020, https://blogs.mercola.com/sites/vitalvotes/archive/2020/12/27/fauci-says-schools-should
-try-to-stay-open.aspx.

40. Ariana Eunjung Cha, Loveday Morris, and Michael Birnbaum, "Covid-19 Death Rates Are Lower Worldwide, But No One Is Sure Whether That's a Blip or a Trend," *Washington*

Post, October 9, 2020, https://www.washingtonpost.com/health/2020/10/09/covid
-mortality-rate-down.

41. Alex Berenson, *Unreported Truths About COVID-19 and Lockdowns* (New Jersey: Bowker, 2020), 20.

42. Centers for Disease Control and Prevention, "CDC 2019 Novel Coronavirus RT-PCR Diagnostic Panel," July 13, 2020 (PDF), https://www.fda.gov/media/134922/download.

43. Barbara Cáceres, "Coronavirus Cases Plummet When PCR Tests Are Adjusted," *Vaccine Reaction*, September 29, 2020, https://thevaccinereaction.org/2020/09/coronavirus-cases -plummet-when-pcr-tests-are-adjusted/; Jon Rappoport, "Smoking Gun: Fauci States COVID Test Has Fatal Flaw; Confession from the 'Beloved' Expert of Experts," *Jon Rappoport's Blog*, November 6, 2020, https://blog.nomorefakenews.com/2020/11/06/smoking-gun-fauci -states-covid-test-has-fatal-flaw; Vincent Racaniello, "COVID-19 with Dr. Anthony Fauci," *This Week in Virology* 641 (July 16, 2020), https://youtu.be/a_Vy6fgaBPE?t=260.

44. Jon Rappoport, "Smoking Gun"; Vincent Racaniello, "COVID-19 with Dr. Anthony Fauci," *This Week in Virology*, July 16, 2020 (video), 4:20, https://www.youtube.com/watch?t=260.

45. Rita Jaafar et al., "Correlation Between 3790 Quantitative Polymerase Chain Reaction– Positives Samples and Positive Cell Cultures, Including 1941 Severe Acute Respiratory Syndrome Coronavirus 2 Isolates," *Clinical Infectious Diseases* ciaa 1491 (September 28, 2020), https://doi.org/10.1093/cid/ciaa1491.

46. Victor Corman et al., "Diagnostic Detection of Wuhan Coronavirus 2019 by Real-Time RT-PCR, January 13, 2020," WHO.int. January 13, 2020 (PDF), https://www.who.int /docs/default-source/coronaviruse/wuhan-virus-assay-v1991527e5122341d99287a1b17c 111902.pdf; Victor M. Corman et al., "Detection of 2019 Novel Coronavirus (2019- nCoV) by Real-Time RT-PCR," *Eurosurveillance* 25, no. 3 (2020): pii 2000045, https:// www.doi.org/10.2807/1560-7917.ES.2020.25.3.2000045.

47. Centers for Disease Control and Prevention, "CDC 2019 Novel Coronavirus RT-PCR Diagnostic Panel," July 13, 2020 (PDF), https://www.fda.gov/media/134922/download.

48. Stacey Lennox, "PREDICTION: Joe Biden Would Manage COVID-19 in One of Two Ways—Both Should Infuriate You," PJ Media, October 27, 2020, https://pjmedia.com /columns/stacey-lennox/2020/10/27/prediction-joe-biden-would-manage-covid-19-in-one -of-two-ways-both-should-infuriate-you-n1092407; "COVID-19: Do We Have a Coronavirus Pandemic, or a PCR Test Pandemic?," Association of American Physicians and Surgeons, October 7, 2020, https://aapsonline.org/covid-19-do-we-have-a-coronavirus -pandemic-or-a-pcr-test-pandemic/.

49. Rappoport, "Smoking Gun."

50. Bernard La Scola et al., "Viral RNA Load as Determined by Cell Culture as a Man- agement Tool for Discharge of SARS-CoV-2 Patients from Infectious Disease Wards," *European Journal of Clinical Microbiology & Infectious Diseases* 39 (2020): 1059–61, https:// doi.org/10.1007/s10096-020-03913-9.

51. T. Jefferson et al., "Viral Cultures for COVID-19 Infectious Potential Assessment— A Systematic Review," *Clinical Infectious Diseases* ciaa 1764, December 3, 2020, https://doi .org/10.1093/cid/ciaa1764.

52. "Every Scary Thing You're Being Told, Depends on the Unreliable PCR Test," YouTube, December 27, 2020 (video), https://www.youtube.com/watch?app=desktop&v=6ny 9nNFHQsY&feature=youtu.be.

53. Florida Health, "Mandatory Reporting of COVID-19 Laboratory Test Results: Reporting of Cycle Threshold Values," December 3, 2020, https://www.flhealthsource.gov/files /Laboratory-Reporting-CT-Values-12032020.pdf.

54. Tyler Durden, "For the First Time, a US State Will Require Disclosure of PCR 'Cycle Threshold' Data in COVID Tests," ZeroHedge, December 7, 2020, https://www.zerohedge .com/medical/first-time-us-state-will-require-disclosure-pcr-test-cycle-data.

55. Pieter Borger et al., "External Peer Review of the RTPCR Test to Detect SARS-CoV-2 Reveals 10 Major Scientific Flaws at the Molecular and Methodological Level: Conse- quences for False Positive Results," Corman-Drosten Review Report, November 27, 2020, https://cormandrostenreview.com/report.

56. Dr. Wolfgang Wodarg and Dr. Michael Yeadon, "Petition/Motion for Administrative/ Regulatory Action Regarding Confirmation of Efficacy End Points and Use of Data Connection with the Following Clinical Trial(s)," Corona Transition, December 1, 2020, https://corona-transition.org/IMG/pdf/wodarg_yeadon_ema_petition_pfizer_trial_final _01dec2020_signed_with_exhibits_geschwa_rzt.pdf.

57. Borger et al., "External Peer Review of the RTPCR Test."

58. Corman et al., "Detection of 2019 Novel Coronavirus (2019-nCoV) by Real-Time RT-PCR."

59. Acu2020.org, "A20 Chief Inspector Michael Fritsch in the Extra-Parliamentary Corona Committee of Inquiry (English Version)," accessed January 20, 2021, https://acu2020 .org/english-versions/; Dr. Reiner Fuellmich, "German Corona Investigative Committee," *Algora* (blog), October 4, 2020, https://www.algora.com/Algora_blog/2020/10/04/german -corona-investigative-committee.

60. Celia Farber, "Ten Fatal Errors: Scientists Attack Paper That Established Global PCR Driven Lockdown," *UncoverDC*, December 3, 2020, https://uncoverdc.com/2020/12/03 /ten-fatal-errors-scientists-attack-paper-that-established-global-pcr-driven-lockdown/.

61. Borger et al., "External Peer Review of the RTPCR Test."

62. Shiyi Cao et al., "Post-Lockdown SARS-CoV-2 Nucleic Acid Screening."

63. "Attorney Dr. Reiner Fuellmich: The Corona Fraud Scandal Must Be Criminally Prosecuted."

64. "CDC 2019-Novel Coronavirus (2019-nCoV) Real-Time RT-PCR Diagnostic Panel," US Food and Drug Administration, revised and updated December 1, 2020, https://www .fda.gov/media/134922/download.

65. "Open Letter from Medical Doctors and Health Professionals to All Belgian Authorities and All Belgian Media," Docs 4 Open Debate, September 5, 2020, https://docs4opendebate .be/en/open-letter.

66. "WHO Information Notice for IVD Users 2020/05," World Health Organization, January 20, 2021, https://www.who.int/news/item/20-01-2021-who-information-notice-for-ivd -users-2020-05.

67. Jamey Keaten, "Biden's US Revives Support for WHO, Reversing Trump Retreat," *AP NEWS*, January 21, 2021, https://apnews.com/article/us-who-support-006ed181e016afa 55d4cea30af236227.

68. "WHO Information Notice for IVD Users." World Health Organization, December 14, 2020, https://web.archive.org/web/20201222013649/https://www.who.int/news/item/14-12 -2020-who-information-notice-for-ivd-users.

69. "WHO Information Notice for IVD Users 2020/05."

70. Meryl Nass, MD, "Shameless Manipulation: Positive PCR Tests Drop after WHO Instructs Vendors to Lower Cycle Thresholds: We Have Been Played Like a Fiddle," *Anthrax Posts* (blog), February 12, 2021, https://anthraxvaccine.blogspot.com/2021/02/positivity-of-pcr-tests-drops-as.html.

71. "US Currently Hospitalized," The COVID Tracking Project, https://covidtracking.com/data/charts/us-currently-hospitalized.

72. Nass, "Shameless Manipulation."

73. "Lord Sumption on the National 'Hysteria' Over Coronavirus," UnHerd, *The Post*, March 30, 2020, https://unherd.com/thepost/lord-sumption-on-the-national-coronavirus-hysteria.

74. Amnesty International, "Biderman's Chart of Coercion," 1994 (PDF), https://www.strath.ac.uk/media/1newwebsite/departmentsubject/socialwork/documents/eshe/Bidermanschartofcoercion.pdf; Center for the Study of Human Rights in the Americas at the University of California at Davis, "Military Training Materials," http://humanrights.ucdavis.edu/projects/the-guantanamo-testimonials-project/testimonies/testimonies-of-the-defense-department/military-training-materials.

Chapter Six: Protecting Yourself from COVID-19

1. Monique Tan, Feng J. He, and Graham A. MacGregor, "Obesity and Covid-19: The Role of the Food Industry," *BMJ* 369 (2020): m2237, https://doi.org/10.1136/bmj.m2237.

2. "Partnership for an Unhealthy Planet," Corporate Accountability, accessed December 15, 2020, https://www.corporateaccountability.org/wp-content/uploads/2020/09/Partnership-for-an-unhealthy-planet.pdf.

3. Gareth Iacobucci, "Food and Soft Drink Industry Has Too Much Influence over US Dietary Guidelines, Report Says," *BMJ* 369 (2020): m1666, https://doi.org/10.1136/bmj.m1666.

4. Sarah Steele et al., "Are Industry-Funded Charities Promoting 'Advocacy-Led Studies' or 'Evidence-Based Science'?: A Case Study of the International Life Sciences Institute," *Globalization and Health* 15, no. 36 (2019), https://doi.org/10.1186/s12992-019-0478-6.

5. "Partnership for an Unhealthy Planet."

6. Anaïs Rico-Campà et al., "Association Between Consumption of Ultra-Processed Foods and All Cause Mortality: SUN Prospective Cohort Study," *BMJ* 365 (2019), https://doi.org/10.1136/bmj.l1949.

7. "Provisional Death Counts for Coronavirus Disease 2019 (COVID-19)," Centers for Disease Control and Prevention, accessed August 26, 2020, https://www.cdc.gov/nchs/nvss/vsrr/covid_weekly/index.htm.

8. "Government Launches Obesity Strategy," BBC News, July 27, 2020, https://www.youtube.com/watch?app=desktop&v=55CrH0fGWFA&feature=youtu.be; Dr. Aseem Malhotra (@DrAseemMalhotra), tweet, "'The government and public health england are ignorant and grossly negligent for not telling the public they need to change their diet now . . . ,'" April 20, 2020, https://twitter.com/DrAseemMalhotra/status/1252253860497948674.

9. Oliver Morrison, "Coronavirus and Obesity: Doctors Take Aim at Food Industry over Poor Diets," FOODnavigator.com, last updated April 27, 2020, https://www.foodnavigator.com/Article/2020/04/22/Coronavirus-and-obesity-Doctors-take-aim-at-food-industry-over-poor-diets.

10. Morrison, "Coronavirus and Obesity."

11. Aseem Malhotra, "Covid 19 and the Elephant in the Room," *European Scientist*, April 16, 2020, https://www.europeanscientist.com/en/article-of-the-week/covid-19-and-the -elephant-in-the-room.

12. Bee Wilson, "How Ultra-Processed Food Took over Your Shopping Basket," *Guardian*, February 12, 2020, https://www.theguardian.com/food/2020/feb/13/how-ultra-processed -food-took-over-your-shopping-basket-brazil-carlos-monteiro.

13. "Interactive Web Tool Maps Food Deserts, Provides Key Data," US Department of Agriculture blog, February 21, 2017, https://www.usda.gov/media/blog/2011/05/03/interactive -web-tool-maps-food-deserts-provides-key-data.

14. Bara El-Kurdi et al., "Mortality from Coronavirus Disease 2019 Increases with Unsaturated Fat and May Be Reduced by Early Calcium and Albumin Supplementation," *Gastroenterology* 159, no. 3 (2020): 1015–18.e4, https://www.doi.org/10.1053/j.gastro.2020.05.057.

15. Andrea Di Francesco et al., "A Time to Fast," *Science* 362, no. 6416 (November 16, 2018): 770–75, https://doi.org/10.1126/science.aau2095.

16. Amy T. Hutchison et al., "Time-Restricted Feeding Improves Glucose Tolerance in Men at Risk for Type 2 Diabetes: A Randomized Crossover Trial," *Obesity*, April 19, 2019, https://doi.org/10.1002/oby.22449.

17. "How to Boost Your Immune System," Harvard Health Publishing, Harvard Medical School, last updated April 6, 2020, https://www.health.harvard.edu/staying-healthy/how -to-boost-your-immune-system; Ruth Sander, "Exercise Boosts Immune Response," *Nursing Older People* 24, no. 6 (June 29, 2012): 11, https://doi.org/10.7748/nop.24.6.11.s11.

18. Josh Barney, "Exercise May Protect Against Deadly Covid-19 Complication, Research Suggests," *UVA Today*, April 15, 2020, https://news.virginia.edu/content/exercise-may -protect-against-deadly-covid-19-complication-research-suggests; University of Virginia Health System, "COVID-19: Exercise May Protect Against Deadly Complication," EurekAlert!, April 15, 2020, https://www.eurekalert.org/pub_releases/2020-04/uovh-cem 041520.php; Zhen Yan and Hanna R. Spaulding, "Extracellular Superoxide Dismutase, A Molecular Transducer of Health Benefits of Exercise," *Redox Biology* 32 (May 2020): 101508, https://doir.org/10.1016/j.redox.2020.101508.

19. Christopher Weyh, Karsten Krüger, and Barbara Strasser, "Physical Activity and Diet Shape the Immune System During Aging," *Nutrients* 12, no. 3 (2020): 622, https://doi.org /10.3390/nu12030622.

20. "Coping with Stress," Centers for Disease Control and Prevention, updated December 11, 2020, https://cdc.gov/coronavirus/2019-ncov/daily-life-coping/managing-stress-anxiety.html.

21. S. K. Agarwal and G. D. Marshall, Jr., "Stress Effects on Immunity and Its Application to Clinical Immunology," *Clinical and Experimental Allergy* 31 (2001): 25–31, https://media .gradebuddy.com/documents/1589333/fcfea000-0fb6-4dde-b786-dda6725fd20c.pdf.

22. Jennifer N. Morey et al., "Current Directions in Stress and Human Immune Function," *Current Opinion in Psychology* 5 (October 2015): 13–17, https://doi.org/10.1016/j.copsyc.2015.03.007.

23. Tobias Esch, Gregory L. Fricchione, and George B. Stefano, "The Therapeutic Use of the Relaxation Response in Stress-Related Diseases," *Medical Science Monitor* 9, no. 2 (February 2003): RA23–34, https://pubmed.ncbi.nlm.nih.gov/12601303.

24. Bruce Barrett et al., "Meditation or Exercise for Preventing Acute Respiratory Infection: A Randomized Controlled Trial," *Annals of Family Medicine* 10, no. 4 (July 2012): 337–46, https://doi.org/10.1370/afm.1376.

25. Harvey W. Kaufman et al., "SARS-CoV-2 Positivity Rates Associated with Circulating 25-Hydroxyvitamin D Levels," *PLoS One* 15 (September 17, 2020): e0239252, https://doi .org/10.1371/journal.pone.0239252.

26. Carlos H. Orces, "Vitamin D Status Among Older Adults Residing in the Littoral and Andes Mountains in Ecuador," *Scientific World Journal* (2015): 545297, https://doi.org /10.1155/2015/545297.

27. "dminder," dminder.ontometrics.com, https://dminder.ontometrics.com.

28. "Are Both Supplemental Magnesium and Vitamin K_2 Combined Important for Vitamin D Levels?," GrassrootsHealth Nutrient Research Institute, accessed December 18, 2020, https://www.grassrootshealth.net/blog/supplemental-magnesium-vitamin-k2-combined -important-vitamin-d-levels/.

29. "Are Both Supplemental Magnesium and Vitamin K_2 Combined Important for Vitamin D Levels?"

30. Alexey V. Polonikov, "Endogenous Deficiency of Glutathione as the Most Likely Cause of Serious Manifestations and Death in Patients with the Novel Coronavirus Infection (COVID-19): A Hypothesis Based on Literature Data and Own Observations," preprint, https://www.researchgate.net/publication/340917045_Endogenous_deficiency_of_glutathione _as_the_most_likely_cause_of_serious_manifestations_and_death_in_patients_with_the _novel_coronavirus_infection_COVID-19_a_hypothesis_based_on_literature_data_and_o.

31. Dr. Joseph Debé, "NAC Is Being Studied in COVID-19. Should You Take It?" *Nutritious Bytes* (blog), April 3, 2020, https://www.drdebe.com/blog/2020/4/2/0txsap858db2lx8l6b 21fultjorb4x; S. De Flora, C. Grassi, and L. Carati, "Attenuation of Influenza-Like Symptomatology and Improvement of Cell-Mediated Immunity with Long-Term N-acetylcysteine Treatment," *European Respiratory Journal* 10 (1997): 1535–41, https://erj .ersjournals.com/content/10/7/1535.long.

32. De Flora, Grassi, and Carati, "Attenuation of Influenza-Like Symptomatology and Improvement of Cell-Mediated Immunity."

33. Vittorio Demicheli et al., "Vaccines for Preventing Influenza in Healthy Adults," *Cochrane Database of Systematic Reviews*, March 13, 2014, updated February 1, 2018, https://doi .org/10.1002/14651858.CD001269.pub5.

34. Adrian R. Martineau et al., "Vitamin D Supplementation to Prevent Acute Respiratory Tract Infections: Systematic Review and Meta-Analysis of Individual Participant Data," *BMJ* 256 (2017): i6583, https://doi.org/10.1136/bmj.i6583.

35. Alexey Polonikov, "Endogenous Deficiency of Glutathione as the Most Likely Cause of Serious Manifestations and Death in COVID-19 Patients," ACS Infectious Diseases 6, no. 7 (2020): 1558–62, https://doi.org/10.1021/acsinfecdis.0c00288.

36. Bin Wang, Tak Yee Aw, and Karen Y. Stokes, "N-acetylcysteine Attenuates Systemic Platelet Activation and Cerebral Vessel Thrombosis in Diabetes," *Redox Biology* 14 (2018): 218–28., https://doi.org/10.1016/j.redox.2017.09.005.

37. Sara Martinez de Lizarrondo, et al., "Potent Thrombolytic Effect of N-Acetylcysteine on Arterial Thrombi," *Circulation* 136, no. 7 (2017): 646-60, https://doi.org/10.1161/ CIRCULATIONAHA.117.027290.

38. Francis L. Poe and Joshua Corn, "N-Acetylcysteine: A Potential Therapeutic Agent for SARS-CoV-2," *Medical Hypotheses* 143 (2020): 109862, https://doi.org/10.1016/j.mehy .2020.109862.

39. "ClinicalTrials.gov," U.S. National Library of Medicine, accessed February 8, 2021, https://clinicaltrials.gov/ct2/results?recrs=&cond=COVID-19&term=NAC&cntry=&state=&city=&dist=.

40. G. A. Eby, D. R. Davis, and W. W. Halcomb, "Reduction in Duration of Common Colds by Zinc Gluconate Lozenges in a Double-Blind Study," *Antimicrobial Agents and Chemotherapy* 25, no. 1 (1984): 20–24, https://doi.org/10.1128/aac.25.1.20.

41. Harri Hemilä, "Zinc Lozenges and the Common Cold: A Meta-Analysis Comparing Zinc Acetate and Zinc Gluconate, and the Role of Zinc Dosage," *JRSM Open* 8, no. 5 (May 2017): 2054270417694291, https://dx.doi.org/10.1177%2F2054270417694291.

42. Zafer Kurugöl, "The Prophylactic and Therapeutic Effectiveness of Zinc Sulphate on Common Cold in Children," *Acta Paediatrica* 95, no. 10 (November 2006): 1175–81, https://doi.org/10.1080/08035250600603024.

43. Dinesh Jothimani et al., "COVID-19: Poor Outcomes in Patients with Zinc Deficiency," *International Journal of Infectious Disease* 100 (November 2020): 343–49, https://dx.doi.org/10.1016%2Fj.ijid.2020.09.014.

44. Artwaan J. W. te Velthuis et al., "Zn^{2+} Inhibits Coronavirus and Arterivirus RNA Polymerase Activity *in Vitro* and Zinc Ionophores Block the Replication of These Viruses in Cell Culture," *PLoS Pathogens* 6 (November 4, 2010): e1001176, https://doi.org/10.1371/journal.ppat.1001176.

45. "Zinc Fact Sheet for Health Professionals," US Department of Health and Human Services, National Institutes of Health, updated July 15, 2020, https://ods.od.nih.gov/factsheets/Zinc-HealthProfessional.

46. Venkataramanujan Srinivasan, PhD, et al., "Melatonin in Septic Shock—Some Recent Concepts," *Journal of Critical Care* 25 (2010): 656.e1–656.e6, https://www.researchgate.net/publication/261798535_melatonin_and_septic_shock_-_some_recent_concepts.

47. Grazyna Swiderska-Kołacz, Jolanta Klusek, and Adam Kołataj, "The Effect of Melatonin on Glutathione and Glutathione Transferase and Glutathione Peroxidase Activities in the Mouse Liver and Kidney In Vivo," *Neuro Endocrinology Letters* 27, no. 3 (June 2006): 365-8, https://pubmed.ncbi.nlm.nih.gov/16816830.

48. Dun-Xian Tan et al., "Melatonin: A Hormone, a Tissue Factor, an Autocoid, a Paracoid, and an Antioxidant Vitamin," *Journal of Pineal Research* 34, no. 1 (December 17, 2002), https://doi.org/10.1034/j.1600-079X.2003.02111.x.

49. Feres José Mocayar Marón et al., "Daily and Seasonal Mitochondrial Protection: Unraveling Common Possible Mechanisms Involving Vitamin D and Melatonin," *Journal of Steroid Biochemistry and Molecular Biology* 199 (May 2020): 105595, https://doi.org/10.1016/j.jsbmb.2020.105595.

50. Dario Acuna-Castroviejo et al., "Melatonin Role in the Mitochondrial Function," *Frontiers in Bioscience* 12 (January 1, 2007): 947–63, https://doi.org/10.2741/2116.

51. Sylvie Tordjman et al., "Melatonin: Pharmacology, Functions and Therapeutic Benefits," *Current Neuropharmacology* 15, no. 3 (April 2017): 434–43, https://doi.org/10.2174/1570159X14666161228122115.

52. Antonio Carrillo-Vico et al., "Melatonin: Buffering the Immune System," *International Journal of Molecular Sciences* 14, no. 4 (April 2013): 8638–83, https://doi.org/10.3390/ijms14048638.

53. Y. Yadi Zhou et al., "A Network Medicine Approach to Investigation and Population-Based Validation of Disease Manifestations and Drug Repurposing for COVID-19," *PLoS Biology* 18, no. 11 (November 6, 2020): e3000970, https://doi.org/10.1371/journal.pbio.3000970.

54. Alpha A Fowler, III, et al., "Effect of Vitamin C Infusion on Organ Failure and Bio-markers of Inflammation and Vascular Injury in Patients with Sepsis and Severe Acute Respiratory Failure: The CITRIS-ALI Randomized Clinical Trial," *JAMA* 322, no. 13 (2019): 1261–70, https://doi.org/10.1001/jama.2019.11825.

55. Ruben Manuel Luciano Colunga Biancatelli et al., "Quercetin and Vitamin: An Experimental, Synergistic Therapy for the Prevention and Treatment of SARS-CoV-2 Related Disease (COVID-19)," *Frontiers in Immunology*, June 19, 2020, https://doi.org/10.3389/fimmu.2020.01451.

56. Patrick Holford et al., "Vitamin C—an Adjunctive Therapy for Respiratory Infection, Sepsis, and COVID-19," *Nutrients* 12, no. 12 (December 7, 2020): 3760, https://www.mdpi.com/2072-6643/12/12/3760/htm.

57. Holford et al., "Vitamin C—an Adjunctive Therapy."

58. Front Line COVID-19 Critical Care Alliance, January 14, 2020, https://covid19criticalcare.com.

59. Paul Marik, MD, "EVMS Critical Care COVID-19 Management Protocol," Eastern Virginia Medical School, August 1, 2020, https://www.evms.edu/media/evms_public/departments/internal_medicine/EVMS_Critical_Care_COVID-19_Protocol.pdf.

60. US National Library of Medicine, "Glucose-6-Phosphate Dehydrogenase Deficiency," MedlinePlus, accessed January 20, 2021, https://ghr.nlm.nih.gov/condition/glucose-6-phosphate-dehydrogenase-deficiency.

61. S. F. Yanuck et al., "Evidence Supporting a Phased Immuno-Physiological Approach to COVID-19 from Prevention Through Recovery," *Integrative Medicine* 19, no. S1 (2020) [Epub ahead of print] (PDF), https://athmjournal.com/covid19/wp-content/uploads/sites/4/2020/05/imcj-19-08.pdf.

62. Ling Yi et al., "Small Molecules Blocking the Entry of Severe Acute Respiratory Syndrome Coronavirus into Host Cells," *Journal of Virology* 78, no. 20 (September 2004): 11334–39, https://doi.org/10.1128/JVI.78.20.11334-11339.2004; Lili Chen et al., "Binding Interaction of Quercetin-3-beta-galactoside and Its Synthetic Derivatives with SARS-CoV 3CL(pro): Structure-Activity Relationship Studies Reveal Salient Pharmacophore Features," *Bioorganic & Medicinal Chemistry* 14, no. 24 (2006): 8295–306, https://doi.org/10.1016/j.bmc.2006.09.014; Nick Taylor-Vaisey, "A Made-in-Canada Solution to the Coronavirus Outbreak," *Maclean's*, February 24, 2020, https://www.macleans.ca/news/canada/a-made-in-canada-solution-to-the-coronavirus-outbreak/.

63. Nicholas Smith and Jeremy C. Smith, "Repurposing Therapeutics for COVID-19: Supercomputer-Based Docking to the SARS-CoV-2 Viral Spike Protein and Viral Spike Protein-Human ACE2 Interface," ChemRxiv preprint, November 3, 2020, https://doi.org/10.26434/chemrxiv.11871402.v4; "Quercetin—a Treatment for Coronavirus?" Greenstarsproject.org, March 27, 2020, https://greenstarsproject.org/2020/03/27/quercetin-a-treatment-for-coronavirus/.

64. Yao Li et al., "Quercetin, Inflammation and Immunity," *Nutrients* 8, no. 3 (March 15, 2016): 167, https://doi.org/10.3390/nu8030167.

65. Yao Li et al., "Quercetin, Inflammation and Immunity."

66. Husam Dabbagh-Bazarbachi et al., "Zinc Ionophore Activity of Quercetin and Epigallocatechin-Gallate: From Heba 1-6 Cells to a Liposome Model," *Journal of Agricultural and Food Chemistry* 62, 32 (July 22, 2014): 8085–93, https://doi.org/10.1021/jf5014633.

67. James J. DiNicolantonio and Mark F. McCarty, "Targeting Casein Kinase 2 with Quercetin or Enzymatically Modified Isoquercitrin as a Strategy for Boosting the Type 1 Interferon Response to Viruses and Promoting Cardiovascular Health," *Medical Hypotheses* 142 (2020): 109800, https://doi.org/10.1016/j.mehy.2020.109800.

68. DiNicolantonio and McCarty, "Targeting Casein Kinase 2 with Quercetin or Enzymatically Modified Isoquercitrin."

69. József Tözser and Szilvia Benkö, "Natural Compounds as Regulators of NLRP3 Inflammasome-Mediated IL-1β Production," *Mediators of Inflammation* 2016 (2016): 5460302, https://doi.org/10.1155/2016/5460302.

70. Ling Yi et al., "Small Molecules Blocking the Entry"; Thi Thanh Hanh Nguyen et al., "Flavonoid-Mediated Inhibition of SARS Coronavirus 3C-Like Protease Expressed in *Pichia pastoris*," *Biotechnology Letters* 34 (2012): 831–38, https://doi.org/10.1007/s10529-011-0845-8; Young Bae Ryu et al., "Biflavonoids from *Torreya nucifera* Displaying SARS-CoV 3CL^{pro} Inhibition," *Bioorganic & Medicinal Chemistry* 18, no. 22 (2010): 7940–47, https://doi.org/10.1016/j.bmc.2010.09.035.

71. Siti Khaerunnisa et al., "Potential Inhibitor of COVID-19 Main Protease (M^{pro}) from Several Medicinal Plant Compounds by Molecular Docking Study," *Preprints* 2020, 2020030226, Preprints.org, March 12, 2020, https://doi.org/10.20944/preprints202003.0226.v1.

72. Paul Marik, MD, "EVMS Critical Care COVID-19 Management Protocol," Eastern Virginia Medical School, August 1, 2020, https://www.evms.edu/media/evms_public/departments/internal_medicine/EVMS_Critical_Care_COVID-19_Protocol.pdf.

73. Hira Shakoor et al., "Be Well: A Potential Role for Vitamin B in COVID-19," *Maturitas* 144 (2021): 108–11, https://doi.org/10.1016/j.maturitas.2020.08.007.

74. Shakoor et al., "Be Well."

75. Dmitry Kats, PhD, MPH, "Sufficient Niacin Supply: The Missing Puzzle Piece to COVID-19, and Beyond?" OSF Preprints, December 29, 2020, https://osf.io/uec3r/.

76. Shakoor et al., "Be Well."

77. Zahra Sheybani et al., "The Rise of Folic Acid in the Management of Respiratory Disease Caused by COVID-19," ChemRxiv preprint, March 30, 2020, https://doi.org/10.26434/chemrxiv.12034980.v1

78. Sheybani et al., "The Rise of Folic Acid in the Management."

79. Vipul Kumar and Manoj Jena, "In Silico Virtual Screening-Based Study of Nutraceuticals Predicts the Therapeutic Potentials of Folic Acid and Its Derivatives Against COVID-19," Research Square, May 26, 2020, https://doi.org/10.21203/rs.3.rs-31775/v1.

80. A. David Smith et al., "Homocysteine-Lowering by B Vitamins Slows the Rate of Accelerated Brain Atrophy in Mild Cognitive Impairment: A Randomized Controlled Trial," *PloS One* 5, no. 9 (September 8, 2010): e12244, https://doi.org/10.1371/journal.pone.0012244.

81. Jane E. Brody, "Vitamin B$_{12}$ as Protection for the Aging Brain," *New York Times*, September 6, 2016, https://www.nytimes.com/2016/09/06/well/mind/vitamin-b12-as-protection-for-the-aging-brain.html.

82. Shakoor et al., "Be Well."

83. Mark F. McCarty and James J DiNicolantonio, "Nutraceuticals Have Potential for Boosting the Type 1 Interferon Response to RNA Viruses Including Influenza and Coronavirus," *Progress in Cardiovascular Diseases* 63, no. 3 (2020): 383–85, https://doi.org/10.1016/j.pcad.2020.02.007.

84. Lionel B. Ivashkiv and Laura T. Donlin, "Regulation of Type I Interferon Responses," *Nature Reviews Immunology* 14, no. 1 (2014): 36–49, https://doi.org/10.1038/nri3581.

85. Anna T. Palamara, "Inhibition of Influenza A Virus Replication by Resveratrol," *Journal of Infectious Diseases* 191, no. 10 (May 15, 2005): 1719–29, https://academic.oup.com/jid/article/191/10/1719/790275.

86. Kai Zhao et al., "Perceiving Nasal Patency Through Mucosal Cooling Rather than Air Temperature or Nasal Resistance," *PLoS One* 6, no. 10 (2011): e24618, https://doi.org/10.1371/journal.pone.0024618.

87. A. V. Arundel et al., "Indirect Health Effects of Relative Humidity in Indoor Environments," *Environmental Health Perspectives* 65 (March 1986): 351–61, https://dx.doi.org/10.1289%2Fehp.8665351.

88. Gordon Lauc et al., "Fighting COVID-19 with Water," *Journal of Global Health* 10, no. 1 (June 2020), http://jogh.org/documents/issue202001/jogh-10-010344.pdf.

89. Eriko Kudo et al., "Low Ambient Humidity Impairs Barrier Function and Innate Resistance Against Influenza Infection," *PNAS* 116, no. 22 (May 28, 2019): 10905–10, https://doi.org/10.1073/pnas.1902840116.

90. J. M. Reiman et al., "Humidity as a Non-Pharmaceutical Intervention for Influenza A," *PLoS One* 13, no. 9 (September 25, 201): e0204337, https://doi.org/10.1371/journal.pone.0204337.

Chapter Seven: Pharmaceutical Failures in the COVID-19 Crisis

1. Patricia J. García, "Corruption in Global Health: The Open Secret," *Lancet* 394, no. 10214 (December 7, 2019): 2119–24, https://doi.org/10.1016/S0140-6736(19)32527-9.

2. García, "Corruption in Global Health."

3. Ron Law, "Rapid Response: WHO Changed Definition of Influenza Pandemic," *BMJ* 2010, no. 340 (June 6, 2010): c2912, https://www.bmj.com/rapid-response/2011/11/02/who-changed-definition-influenza-pandemic; World Health Organization. "Epidemic and Pandemic Alert and Response," Wayback Machine, archived May 11, 2009 (PDF), http://whale.to/vaccine/WHO1.pdf.

4. World Health Organization, "Pandemic Preparedness," Wayback Machine, archived September 2, 2009 (PDF), http://whale.to/vaccine/WHO2.pdf.

5. Merritt, "SARS-CoV2 and the Rise of Medical Technocracy"; D. G. Rancourt, "All-Cause Mortality During COVID-19: No Plague and a Likely Signature of Mass Homicide by Government Response," *Technical Report*, June 2020, https://www.doi.org/10.13140/RG.2.24350.77125; Yanni Gu, "A Closer Look at US Deaths Due to COVID-19," *Johns Hopkins News-Letter* web archive, November 22, 2020, https://web.archive.org/web/20201126163323/https:/www.jhunewsletter.com/article/2020/11/a-closer-look-at-u-s-deaths-due-to-covid-19.

6. Lisa M. Krieger, "Stanford Researcher Says Coronavirus Isn't as Fatal as We Thought; Critics Say He's Missing the Point," *Mercury News*, May 20, 2020 (archived), https://archive.is/IWWCC; Justin Blackburn, PhD, et al., "Infection Fatality Ratios for COVID-19 Among Noninstitutionalized Persons 12 and Older: Results of a Random-Sample Prevalence Study," *Annals of Internal Medicine*, 2020, https://www.acpjournals.org/doi/10.7326/M20-5352; Edwin Mora, "Doctor to Senators: Coronavirus Fatality Rate 10 to 40x Lower than Estimates That Led to Lockdowns," *Breitbart*, May 7, 2020, https://www.breitbart.com/politics/2020/05/07/doctor-to-senators-coronavirus-fatality-rate-10-to-40x-lower-than-estimates-that-led-to-lockdowns/; Scott W. Atlas, MD, "How to Re-Open Society Using Evidence, Medical Science, and Logic,"

US Senate testimony, May 6, 2020 (PDF), https://www.hsgac.senate.gov/imo/media/doc /Testimony-Atlas-2020-05-06.pdf; John P. A. Ioannidis, MD, DSc., US Senate testimony, May 6, 2020, https://www.hsgac.senate.gov/imo/media/doc/Testimony-Ioannidis-2020-05-06.pdf.

7. John P. A. Ioannidis, Cathrine Axfors, and Despina G. Contopoulos-Ioannidis, "Population-Level COVID-19 Mortality Risk for Non-Elderly Individuals Overall and for Non-Elderly Individuals Without Underlying Diseases in Pandemic Epicenters," medRxiv preprint, May 5, 2020, https://doi.org/10.1101/2020.04.05.20054361; John P. A. Ioannidis, Cathrine Axfors, and Despina G. Contopoulos-Ioannidis, "Population-Level COVID-19 Mortality Risk for Non-Elderly Individuals Overall and for Non-Elderly Individuals Without Underlying Diseases in Pandemic Epicenters," *Environmental Research* 188 (September 2020): 109890, https://doi.org/10.1016/j.envres.2020.109890.

8. Jeffrey A. Tucker, "WHO Deletes Naturally Acquired Immunity from Its Website," American Institute for Economic Research, December 23, 2020, https://www.aier.org/article/who -deletes-naturally-acquired-immunity-from-its-website/.

9. Tucker, "WHO Deletes Naturally Acquired Immunity from Its Website."

10. Carolyn Dean, *Death by Modern Medicine* (Matrix Verde Media, 2005).

11. Barbara Starfield, "Is US Health Really the Best in the World?," *JAMA* 284 no. 4. (July 26, 2000): 483–85, https://doi.org/10.1001/jama.284.4.483.

12. Martin A. Makary, MA, and Michael Daniel. "Medical Error—The Third Leading Cause of Death in the US," *BMJ* 353 (May 3, 2016): i2139, https://doi.org/10.1136/bmj.i2139.

13. John C. Peters et al., "The Effects of Water and Non-Nutritive Sweetened Beverages on Weight Loss During a 12-Week Weight Loss Treatment Program," *Obesity* 22, no. 6 (June 2014): 1415–21, https://doi.org/10.1002/oby.20737; William Hudson, "Diet Soda Helps Weight Loss, Industry-Funded Study Finds," CNN, May 27, 2014, https://www.cnn.com /2014/05/27/health/diet-soda-weight-loss/index.html.

14. Centers for Disease Control and Prevention, "Disclosure," CDC.gov, accessed January 22, 2021, https://www.cdc.gov/mmwr/cme/serial_conted.html.

15. "US Right to Know Petition to the CDC," November 5, 2019 (PDF): 3, https://usrtk.org /wp-content/uploads/2019/11/Petition-to-CDC-re-Disclaimers.pdf.

16. "US Right to Know Petition to the CDC"; Gary Ruskin, "Groups to CDC: Stop Falsely Claiming Not to Accept Corporate Money," US Right to Know, November 5, 2019, https:// usrtk.org/news-releases/groups-to-cdc-stop-falsely-claiming-not-to-accept-corporate-money/.

17. Chelsea Bard and Lindsey Mills, "Maine Ballot Re-Sparks Vaccination Exemptions Debate," News Center Maine, February 4, 2020, https://www.newscentermaine.com/article /news/politics/maine-ballot-re-sparks-vaccination-exemptions-debate/97-4d9dcc72-2d0d -4575-9c26-da3a685b6d51.

18. Andrew Ward, "Vaccines Are Among Big Pharma's Best Selling Products," *Financial Times*, April 24, 2016, https://www.ft.com/content/93374f4a-e538-11e5-a09b-1f8b0d268c39.

19. Markets and Markets, "Vaccines Market by Technology (Live, Toxoid, Recombinant), Disease (Pneumococcal, Influenza, DTP, Rotavirus, TT, Polio, MMR, Varicella, Dengue, TB, Shingles, Rabies), Route (IM, SC, ID, Oral), Patient (Pediatric, Adult), Type—Global Forecast to 2024," accessed January 22, 2021, https://www.marketsandmarkets.com/Market -Reports/vaccine-technologies-market-1155.html.

20. Bret Stephens, "The Story of Remdesivir," *New York Times*, April 17, 2020, https://www .nytimes.com/2020/04/17/opinion/remdesivir-coronavirus.html.

21. Stephens, "The Story of Remdesivir."

22. Sydney Lupkin, "Remdesivir Priced at More than $3,100 for a Course of Treatment," NPR, June 29, 2020, https://www.npr.org/sections/health-shots/2020/06/29/884648842 /remdesivir-priced-at-more-than-3-100-for-a-course-of-treatment.

23. Elizabeth Woodworth, "Remdesivir for Covid-19: $1.6 Billion for a 'Modestly Beneficial' Drug?," *Global Research*, August 27, 2020, https://www.globalresearch.ca/remdesivir-covid -19-1-6-billion-modestly-beneficial-drug/5717690.

24. Alexa Lorenzo et al., "Florida Seeing 'Explosion' in COVID-19 Cases Among Younger Residents, but Patients Less Sick," WFTV 9, updated June 23, 2020, https://www.wftv .com/news/florida/watch-gov-desantis-speak-orlando-hospital-about-covid-19-1230-pm /UCJ3VAS7ZJDNJKR6KHWZXLCH3E/.

25. John H. Beigel et al., "Remdesivir for the Treatment of Covid-19—Final Report," *New England Journal of Medicine*, 383 (November 5, 2020): 1813–26, https://doi.org/10.1056 /NEJMoa2007764.

26. National Institute of Allergy and Infectious Diseases (NIAID), "Adaptive COVID-19 Treatment Trial (ACTT)," ClinicalTrials.gov, first posted February 21, 2020, last update December 9, 2020, https://clinicaltrials.gov/ct2/show/NCT04280705.

27. FDA News Release, "Coronavirus (COVID-19) Update: FDA Issues Emergency Use Authorization for Potential COVID-19 Treatment," FDA.gov, May 1, 2020, https://www .fda.gov/news-events/press-announcements/coronavirus-covid-19-update-fda-issues -emergency-use-authorization-potential-covid-19-treatment.

28. Yeming Wang et al., "Remdesivir in Adults with Severe COVID-19: A Randomised, Double-Blind, Placebo-Controlled, Multicentre Trial," *Lancet* 395 (2020): 1569–78, https://doi.org/10.1016/S0140-6736(20)31022-9.

29. Marie Dubert et al., "Case Report Study of the First Five COVID-19 Patients Treated with Remdesivir in France," *International Journal of Infectious Diseases* 98 (2020): 290–93, https://doi.org/10.1016/j.ijid.2020.06.093.

30. Yeming Wang et al., "Remdesivir in Adults with Severe COVID-19."

31. US Food and Drug Administration, "Coronavirus Disease 2019 (COVID-19) Resources for Health Professionals," FDA.gov, November 16, 2020, accessed January 22, 2021, https://www.fda.gov/health-professionals/coronavirus-disease-2019-covid-19-resources -health-professionals#testing.

32. FLCCC Alliance, "MATH+ Hospital Treatment Protocol for Covid-19," July 14, 2020, https://covid19criticalcare.com/wp-content/uploads/2020/04/MATHTreatmentProtocol.pdf.

33. Leon Caly et al., "The FDA-Approved Drug Ivermectin Inhibits the Replication of SARS-CoV-2 *in Vitro*," *Antiviral Research* 178 (June 2020): 104787, https://doi.org/10.1016/j .antiviral.2020.104787.

34. FLCCC Alliance, "I-MASK+ Protocol—Downloads and Translations," accessed January 22, 2021, https://covid19criticalcare.com/i-mask-prophylaxis-treatment-protocol/i-mask -protocol-translations/.

35. FLCCC Alliance, "MATH+ Hospital Protocol—Downloads and Translations," accessed January 22, 2021, https://covid19criticalcare.com/math-hospital-treatment/pdf -translations/.

36. Richardson et al., "Presenting Characteristics, Comorbidities, and Outcomes Among 5700 Patients."

37. Joyce Kamen, "The MATH+ Protocol Will Likely Have the Most Dramatic Impact on Survival of Critically Ill Covid19 Patients Worldwide," Medium.com, June 16, 2020, https://joyce-kamen.medium.com/the-math-protocol-will-have-the-most-dramatic-impact-on-survival-of-critically-ill-covid19-35689f7ce16f.

38. FLCCC Alliance, "Front Line COVID-19 Critical Care Alliance," December 8, 2020, https://covid19criticalcare.com/.

39. Swiss Policy Research, "Covid-19: WHO-Sponsored Preliminary Review Indicates Ivermectin Effectiveness," December 31, 2020, https://swprs.org/who-preliminary-review-confirms-Ivermectin-effectiveness/.

40. FLCCC Alliance, "One Page Summary of the Clinical Trials Evidence for Ivermectin in COVID-19," as of January 11, 2021 (PDF), https://covid19criticalcare.com/wp-content/uploads/2020/12/One-Page-Summary-of-the-Clinical-Trials-Evidence-for-Ivermectin-in-COVID-19.pdf.

41. "Ivermectin Meta-Analysis by Dr. Andrew Hill," YouTube, December 27, 2020 (video), https://www.youtube.com/watch?v=yOAh7GtvcOs&feature=emb_logo.

42. US FDA, "FAQ: COVID-19 and Ivermectin Intended for Animals," December 16, 2020, https://www.fda.gov/animal-veterinary/product-safety-information/faq-covid-19-and-ivermectin-intended-animals.

43. FLCCC Alliance, "One Page Summary of the Clinical Trials Evidence for Ivermectin."

44. FLCCC Alliance, "FLCCC Alliance Invited to the National Institutes of Health (NIH) COVID-19 Treatment Guidelines Panel to Present Latest Data on Ivermectin," January 7, 2020 (PDF), https://covid19criticalcare.com/wp-content/uploads/2021/01/FLCCC-PressRelease-NIH-C19-Panel-FollowUp-Jan7-2021.pdf.

45. FLCCC Alliance, "NIH Revises Treatment Guidelines for Ivermectin for the Treatment of COVID-19," January 15, 2021 (PDF), https://covid19criticalcare.com/wp-content/uploads/2021/01/FLCCC-PressRelease-NIH-Ivermectin-in-C19-Recommendation-Change-Jan15.2021-final.pdf.

46. FLCCC Alliance, "NIH Revises Treatment Guidelines for Ivermectin."

47. Matthieu Million et al., "Early Treatment of COVID-19 Patients with Hydroxychloroquine and Azithromycin: A Retrospective Analysis of 1061 Cases in Marseille, France," *Travel Medicine and Infectious Disease* 35 (2020): 101738, https://doi.org/10.1016/j.tmaid.2020.101738; Scott Sayare, "He Was a Science Star. Then He Promoted a Questionable Cure for Covid-19," *New York Times Magazine*, May 12, 2020, https://www.nytimes.com/2020/05/12/magazine/didier-raoult-hydroxychloroquine.html.

48. Phulen Sarma et al., "Virological and Clinical Cure in COVID-19 Patients Treated with Hydroxychloroquine: A Systematic Review and Meta-Analysis," *Journal of Medical Virology* 92, no. 7 (2020): 776–85, https://doi.org/10.1002/jmv.25898.

49. Robert F. Service, "Would-Be Coronavirus Drugs Are Cheap to Make," *Science*, April 10, 2020, https://www.sciencemag.org/news/2020/04/would-be-coronavirus-drugs-are-cheap-make.

50. "Hydroxychloroquine," GoodRx, accessed January 22, 2021, https://www.goodrx.com/hydroxychloroquine.

51. Bill Gates, "What You Need to Know About the COVID-19 Vaccine," *GatesNotes* (blog), April 30, 2020, https://www.gatesnotes.com/Health/What-you-need-to-know-about-the-COVID-19-vaccine.

52. Roland Derwand et al., "COVID-19 Outpatients: Early Risk-Stratified Treatment with Zinc Plus Low-Dose Hydroxychloroquine and Azithromycin: A Retrospective Case Series Study," *International Journal of Antimicrobial Agents* 56, no. 6 (2020): 106214, https://doi.org/10.1016/j.ijantimicag.2020.106214.

53. Harvey A. Risch, MD, PhD, "The Key to Defeating COVID-19 Already Exists. We Need to Start Using It," *Newsweek*, July 23, 2020, https://www.newsweek.com/key-defeating-covid-19-already-exists-we-need-start-using-it-opinion-1519535?amp=1&__twitter_impression=true.

54. Martin J. Vincent et al., "Chloroquine Is a Potent Inhibitor of SARS Coronavirus Infection and Spread," *Virology Journal* 2, no. 69 (August 22, 2005), https://doi.org/10.1186/1743-422X-2-69.

55. Eng Eong Ooi et al., "In Vitro Inhibition of Human Influenza A Virus Replication by Chloroquine," *Virology Journal* 3, no. 39 (May 29, 2006), https://doi.org/10.1186/1743-422X-3-39.

56. Meryl Nass, MD, "WHO 'Solidarity' and U.K. 'Recovery' Clinical Trials of Hydroxychloroquine Using Potentially Fatal Doses," *Age of Autism*, https://www.ageofautism.com/2020/06/who-solidarity-and-uk-recovery-clinical-trials-of-hydroxychloroquine-using-potentially-fatal-doses.html.

57. "Hydroxychloroquine: Drug Information," UpToDate, accessed July 6, 2020, https://www.uptodate.com/contents/hydroxychloroquine-drug-information.

58. World Health Organization, "'Solidarity' Clinical Trial for COVID-19 Treatments," accessed July 6, 2020, https://www.who.int/emergencies/diseases/novel-coronavirus-2019/global-research-on-novel-coronavirus-2019-ncov/solidarity-clinical-trial-for-covid-19-treatments.

59. "Swiss Protocol for COVID—Quercetin and Zinc," Editorials 360, August 20, 2020, https://www.editorials360.com/2020/08/20/swiss-protocol-for-covid-quercetin-and-zinc/.

Chapter Eight: Successful Protocols Suppressed

1. Berkeley Lovelace, Jr., "Pfizer Says Final Data Analysis Shows Covid Vaccine Is 95% Effective, Plans to Submit to FDA in Days," CNBC, November 18, 2020, https://www.cnbc.com/2020/11/18/coronavirus-pfizer-vaccine-is-95percent-effective-plans-to-submit-to-fda-in-days.html; Courtenay Brown, "Stock Market Rises After Pfizer Coronavirus Vaccine News," Axios, November 19, 2020, https://www.axios.com/stock-market-pfizer-coronavirus-vaccine-c3c131d7-b46f-4df0-94c9-503d1dc906df.html; Joe Palca, "Pfizer Says Experimental COVID-19 Vaccine Is More than 90% Effective," NPR, November 9, 2020. https://www.npr.org/sections/health-shots/2020/11/09/933006651/pfizer-says-experimental-covid-19-vaccine-is-more-than-90-effective.

2. Joe Palca, "Moderna's COVID-19 Vaccine Shines in Clinical Trial," NPR, November 16, 2020, https://www.npr.org/sections/health-shots/2020/11/16/935239294/modernas-covid-19-vaccine-shines-in-clinical-trial.

3. Gilbert Berdine, MD, "What the Covid Vaccine Hype Fails to Mention," Mises Wire, November 24, 2020, https://mises.org/wire/what-covid-vaccine-hype-fails-mention.

4. Allen S. Cunningham, MD, November 13, 2020, comment on Elisabeth Mahase, "Covid-19: Vaccine Candidate May Be More than 90% Effective, Interim Results Indicate," *BMJ* 2020, no. 371 (2020): m4347, https://doi.org/10.1136/bmj.m4347.

5. Berdine, "What the Covid Vaccine Hype Fails to Mention."

6. Peter Doshi, "Pfizer and Moderna's 95% Effective Vaccines—Let's Be Cautious and First See the Full Data," *BMJ Opinion*, November 26, 2020, https://blogs.bmj.com/bmj/2020/11/26/peter-doshi-pfizer-and-modernas-95-effective-vaccines-lets-be-cautious-and-first-see-the-full-data/.

7. Doshi, "Pfizer and Moderna's 95% Effective Vaccines."

8. Peter Doshi, "Will Covid-19 Vaccines Save Lives? Current Trials Aren't Designed to Tell Us," *BMJ* 2020, no. 371 (October 21, 2020), https://doi.org/10.1136/bmj.m4037.

9. Doshi, "Will Covid-19 Vaccines Save Lives?"

10. Eyrun Thune, "Modified RNA Has a Direct Effect on DNA," Phys.org, January 29, 2020, https://phys.org/news/2020-01-rna-effect-dna.html.

11. Thune, "Modified RNA Has a Direct Effect on DNA."

12. Damian Garde, "Lavishly Funded Moderna Hits Safety Problems in Bold Bid to Revolutionize Medicine," Stat News, January 10, 2017, https://www.statnews.com/2017/01/10/moderna-trouble-mrna/.

13. James Odell, OMD, ND, L.Ac., "COVID-19 mRNA Vaccines," Bioregulatory Medicine Institute, December 28, 2020, https://www.biologicalmedicineinstitute.com/post/covid-19-mrna-vaccines.

14. Autoimmune Registry, "Estimates of Prevalence for Autoimmune Disease," accessed January 22, 2021, https://www.autoimmuneregistry.org/autoimmune-statistics.

15. Odell, "COVID-19 mRNA Vaccines."

16. Timothy Cardozo and Ronald Veazey, "Informed Consent Disclosure to Vaccine Trial Subjects of Risk of COVID-19 Vaccines Worsening Clinical Disease," *International Journal of Clinical Practice*, October 28, 2020, https://doi.org/10.1111/ijcp.13795.

17. Lisa A. Jackson et al., "An mRNA Vaccine Against SARS-CoV-2—Preliminary Report," *New England Journal of Medicine* 383, no. 20 (2020): 1920–31, https://doi.org/10.1056/NEJMoa2022483.

18. Robert F. Kennedy, Jr., "Catastrophe: 20% of Human Test Subjects Severely Injured from Gates-Fauci Coronavirus Vaccine by Moderna," May 20, 2020, https://fort-russ.com/2020/05/catastrophe-20-of-human-test-subjects-severely-injured-from-gates-fauci-coronavirus-vaccine-by-moderna/.

19. Doshi, "Pfizer and Moderna's 95% Effective Vaccines."

20. Norbert Pardi et al., "mRNA Vaccines—a New Era in Vaccinology," *Nature Reviews Drug Discovery* 2018, no. 17 (January 12, 2018): 261–79, https://www.nature.com/articles/nrd.2017.243.

21. Sissi Cao, "Here Are All the Side Effects of Every Top COVID-19 Vaccine in US," *Observer*, October 20, 2020, https://observer.com/2020/10/vaccine-side-effects-moderna-pfizer-johnson-astrazeneca/.

22. Haley Nelson, Facebook post, December 30, 2020, https://www.facebook.com/photo.php?fbid=10219326599539838&set=p.10219326599539838&type=3; Tara Sekikawa, Facebook post, December 27, 2020, https://www.facebook.com/photo?fbid=10218204338126951&set=a.1290324145245.

23. Karl Dunkin case, Facebook post, January 5, 2021, https://www.facebook.com/marcellaterry/posts/10225204405125047.

24. "Boston Doctor Says He Almost Had to Be INTUBATED After Suffering Severe Allergic Reaction from Moderna Covid Vaccine," RT, December 26, 2020, https://www.rt.com/usa/510775-moderna-covid-vaccine-allergic-reaction/; Children's Health Defense Team, "FDA Investigates Allergic Reactions to Pfizer COVID Vaccine After More Healthcare Workers Hospitalized," *Defender*, December 21, 2020, https://childrenshealthdefense.org/defender/fda-investigates-reactions-pfizer-covid-vaccine-healthcare-workers-hospitalized/?utm_source=salsa&eType=EmailBlastContent&eId=8c0edf71-f718-4f0d-ae2a-84905c

9c8919; Thomas Clark, MD, MPH, "Anaphylaxis Following m-RNA COVID-19 Vaccine Receipt," CDC.gov, December 19, 2020, https://www.cdc.gov/vaccines/acip/meetings /downloads/slides-2020-12/slides-12-19/05-COVID-CLARK.pdf.

25. Children's Health Defense Team, "Fauci: COVID Vaccines Appear Less Effective Against Some New Strains + More," *Defender*, January 12, 2021, https://childrenshealthdefense .org/defender/covid-19-vaccine-news/?utm_source=salsa&eType=EmailBlastContent&eId =62360bc6-a144-49b4-8b74-803793be13fc.

26. Shawn Skelton, Facebook post, January 7, 2021, https://www.facebook.com/shawn.skelton .73/posts/403541337597874; Brant Griner, Facebook post, January 10, 2021, https://www .facebook.com/brant.griner.7/posts/899042044166409; WION Web Team, "Mexican Doctor Admitted to ICU After Receiving Pfizer Covid-19 Vaccine," WioNews, January 2, 2021, https://www.wionews.com/world/mexican-doctor-admitted-to-icu-after-receiving -pfizer-covid-19-vaccine-354093.

27. Alanna Tonge-Jelley, Facebook post, January 9, 2021, https://www.facebook.com/permalink .php?story_fbid=2749373985391622&id=100009571428119.

28. Shivali Best, "Covid Vaccine: Four Pfizer Trial Participants Developed Facial Paralysis, FDA Says," *Mirror*, December 11, 2020, https://www.mirror.co.uk/science/covid-vaccine -four-pfizer-trial-23151047.

29. Sophie Bateman, "Coronavirus Vaccine Patient 'Dies Five Days After Receiving Pfizer Jab,'" *Daily Star*, December 30, 2020, https://www.dailystar.co.uk/news/world-news /breaking-coronavirus-vaccine-patient-dies-23239055; "Health Authorities on Alert After Nurse DIES Following Vaccination with Pfizer's Covid-19 Shot in Portugal," RT, January 4, 2021, https://www.rt.com/news/511524-portuguese-nurse-dies-pfizer-vaccine/; Children's Health Defense Team, "'Perfectly Healthy' Florida Doctor Dies Weeks After Getting Pfizer COVID Vaccine," *Defender*, January 7, 2021, https://childrenshealthdefense .org/defender/healthy-florida-doctor-dies-after-pfizer-covid-vaccine/; Zachary Stieber, "55 People Have Died in US After Receiving COVID-19 Vaccines: Reporting System," *Epoch Times*, January 17, 2021, https://www.theepochtimes.com/55-people-died-in-us-after -receiving-covid-19-vaccines-reporting-system_3659152.html.

30. Clark, "Anaphylaxis Following m-RNA COVID-19 Vaccine Receipt."

31. Centers for Disease Control and Prevention, "COVID-19 Vaccines and Allergic Reactions," CDC.gov, https://www.cdc.gov/coronavirus/2019-ncov/vaccines/safety/allergic-reaction.html.

32. William A. Haseltine, "Covid-19 Vaccine Protocols Reveal That Trials Are Designed to Succeed," *Forbes*, September 23, 2020, https://www.forbes.com/sites/williamhaseltine/2020/09/23 /covid-19-vaccine-protocols-reveal-that-trials-are-designed-to-succeed/?sh=2212afc25247.

33. Danuta M. Skowronski et al., "Association Between the 2008–09 Seasonal Influenza Vaccine and Pandemic H1N1 Illness During Spring–Summer 2009: Four Observational Studies from Canada," *PLoS Medicine*, April 6, 2010, https://doi.org/10.1371/journal.pmed.1000258; Maryn McKenna, "New Canadian Studies Suggest Seasonal Flu Shot Increased H1N1 Risk," CIDRAP, April 6, 2010, https://www.cidrap.umn.edu/news-perspective/2010/04 /new-canadian-studies-suggest-seasonal-flu-shot-increased-h1n1-risk.

34. Ed Susman, "Ferrets Keep Flu Vaccine/H1N1 Pot Boiling," *MedPage Today*, September 9, 2010, https://www.medpagetoday.org/meetingcoverage/icaac/34674?vpass=1.

35. Annie Guest, "Vaccines May Have Increased Swine Flu Risk," March 4, 2011, https:// www.abc.net.au/news/2011-03-04/vaccines-may-have-increased-swine-flu-risk/1967508.

36. Centers for Disease Control and Prevention, "Human Coronavirus Types," CDC.gov, accessed January 22, 2021, https://www.cdc.gov/coronavirus/types.html.

37. Greg G. Wolff, "Influenza Vaccination and Respiratory Virus Interference Among Department of Defense Personnel During the 2017–2018 Influenza Season," *Vaccine* 38, no. 2 (January 10, 2020): 350–54, https://doi.org/10.1016/j.vaccine.2019.10.005; Michael Murray, ND, "Does the Flu Shot Increase COVID-19 Risk (YES!) and Other Interesting Questions," DoctorMurray.com, accessed January 22, 2021, https://doctormurray.com/does-the-flu-shot-increase-covid-19-risk/.

38. Wolff, "Influenza Vaccination and Respiratory Virus Interference,", results and Table 5.

39. American Lung Association, "Human Metapneumovirus (hMPV) Symptoms and Diagnosis," accessed December 20, 2021, https://www.lung.org/lung-health-diseases/lung-disease-lookup/human-metapneumovirus-hmpv/symptoms-diagnosis.

40. Dr. Joseph Mercola, "Vaccine Debate—Kennedy Jr. vs Dershowitz," Mercola.com, August 22, 2020, https://articles.mercola.com/sites/articles/archive/2020/08/22/the-great-vaccine-debate.aspx.

41. Mercola, "Vaccine Debate."

42. Health and Human Services Department, "Declaration Under the Public Readiness and Emergency Preparedness Act for Medical Countermeasures Against COVID-19," *Federal Register*, March 17, 2020, https://www.federalregister.gov/documents/2020/03/17/2020-05484/declaration-under-the-public-readiness-and-emergency-preparedness-act-for-medical-countermeasures.

43. Jon Rappoport, "Exposed: There's a New Federal Court to Handle All the Expected COVID Vaccine Injury Claims," September 22, 2020, https://www.naturalblaze.com/2020/09/exposed-theres-a-new-federal-court-to-handle-all-the-expected-covid-vaccine-injury-claims.html.

44. Rappoport, "Exposed."

45. Justin Blackburn, PhD, "Infection Fatality Ratios for COVID-19 Among Noninstitution-alized Persons 12 and Older: Results of a Random-Sample Prevalence Study," *Annals of Internal Medicine*, January 2021, https://doi.org/10.7326/M20-5352.

46. Rancourt, "All-Cause Mortality During COVID-19"; Merritt, "SARS-CoV-2 and the Rise of Medical Technocracy."

47. Shiyi Cao et al., "Post-Lockdown SARS-CoV-2 Nucleic Acid Screening."

48. Khan Academy, "Adaptive Immunity," accessed January 22, 2021, https://www.khanacademy.org/test-prep/mcat/organ-systems/the-immune-system/a/adaptive-immunity.

49. Alba Grifoni et al., "Targets of T Cell Responses to SARS-CoV-2 Coronavirus in Humans with COVID-19 Disease and Unexposed Individuals," *Cell* 181, no. 7 (2020): 1489–1501. e15, https://doi.org/10.1016/j.cell.2020.05.015; Jason Douglas, "Before Catching Coronavirus, Some People's Immune Systems Are Already Primed to Fight It," *Wall Street Journal*, June 12, 2020 (archived), archive.is/b4UZq.

50. Annika Nelde et al., "SARS-CoV-2-Derived Peptides Define Heterologous and COVID-19-Induced T Cell Recognition," *Nature Immunology* 22 (September 30, 2020): 74–85, https://www.nature.com/articles/s41590-020-00808-x.

51. Anchi Wu, "Interference Between Rhinovirus and Influenza A Virus: A Clinical Data Analysis and Experimental Infection Study," *Lancet Microbe* 1, no. 6 (September 4, 2020): e254–62, https://doi.org/10.1016/S2666-5247(20)30114-2; Brian B. Dunleavy, "Study: Common Cold May Help Prevent Flu, Perhaps COVID-19," UPI, September 4, 2020,

https://www.upi.com/Health_News/2020/09/04/Study-Common-cold-may-help-prevent
-flu-perhaps-COVID-19/7341599247443/.

52. Anchi Wu, "Interference Between Rhinovirus and Influenza A Virus"; Dunleavy, "Study: Common Cold May Help Prevent Flu."

53. Nina Le Bert et al., "SARS-CoV-2-Specific T Cell Immunity in Cases of COVID-19 and SARS, and Uninfected Controls," *Nature* 584, no. 7821 (2020): 457–62, https://doi.org /10.1038/s41586-020-2550-z; Beezy Marsh, "Can a Cold Give You Coronavirus Immunity? Some Forms of Common Respiratory Illness Might Help Build Protection from Covid-19 . . . and It Could Last Up to 17 YEARS, Scientists Say," *Daily Mail*, June 11, 2020, https:// www.dailymail.co.uk/news/article-8412807/Can-cold-coronavirus-immunity.html; Hannah C., "Some Forms of Common Cold May Give COVID-19 Immunity Lasting Up to 17 Years, New Research Suggests," June 12, 2020, https://www.sciencetimes.com /articles/26038/20200612/common-cold-give-covid-19-immunity-lasting-up-17-years.htm.

54. Takya Sekine et al., "Robust T Cell Immunity in Convalescent Individuals with Asymp-tomatic or Mild COVID-19," *Cell* 183, no. 1 (2020): 158–68.e14, https://doi.org/10.1016 /j.cell.2020.08.017.

55. Sekine et al., "Robust T Cell Immunity in Convalescent Individuals."

56. Freddie Sayers, "Karl Friston: Up to 80% Not Even Susceptible to Covid-19," Unherd.com, June 4, 2020, https://unherd.com/2020/06/karl-friston-up-to-80-not-even-susceptible-to -covid-19/.

57. Apoorva Mandavilli, "What If 'Herd Immunity' Is Closer than Scientists Thought?," *New York Times*, August 17, 2020, https://www.nytimes.com/2020/08/17/health/coronavirus -herd-immunity.html.

58. Max Fisher, "R0, the Messy Metric That May Soon Shape Our Lives, Explained," *New York Times*, April 23, 2020, https://www.nytimes.com/2020/04/23/world/europe/coronavirus -R0-explainer.html.

59. Fisher, "R0, the Messy Metric."

60. P. V. Brennan and L. P. Brennan, "Susceptibility-Adjusted Herd Immunity Threshold Model and Potential R_0 Distribution Fitting the Observed Covid-19 Data in Stockholm," medRxiv preprint, May 22, 2020 (PDF), https://doi.org/10.1102/2020.05.19.20104596.

61. M. Gabriela M. Gomes et al., "Individual Variation in Susceptibility or Exposure to SARS-CoV-2 Lowers the Herd Immunity Threshold (PDF)," medRxiv preprint, May 21, 2020, https://www.medrxiv.org/content/10.1101/2020.04.27.20081893v3.full.pdf; J. B. Handley, "Second Wave? Not Even Close," *Off-Guardian*, July 7, 2020, https://off-guardian.org /2020/07/07/second-wave-not-even-close/.

62. Andrew Bostom, "UPDATED—Educating Dr. Fauci on Herd Immunity and Covid-19: Completing What Rand Paul Began," Andrewbostom.org, September 28, 2020, https:// www.andrewbostom.org/2020/09/educating-dr-fauci-on-herd-immunity-and-covid-19 -completing-what-rand-paul-began/.

63. Shmuel Safra, Yaron Oz, and Ittai Rubinstein, "Heterogeneity and Superspreading Effect on Herd Immunity," medRxiv preprint, September 10, 2020, https://doi.org/10.1101/2020 .09.06.20189290.

64. Ricardo Aguas et al., "Herd Immunity Thresholds for SARS-CoV-2 Estimated from Unfolding Epidemics." medRxiv preprint, August 31, 2020, https://doi.org/10.1101/2020 .07.23.20160762.

65. Brennan and Brennan, "Susceptibility-Adjusted Herd Immunity Threshold Model."

66. Tom Britton, Frank Ball, and Pieter Trapman, "The Disease-Induced Herd Immunity Level for Covid-19 Is Substantially Lower than the Classical Herd Immunity Level," Cornell University arXiv.org, May 6, 2020, https://arxiv.org/abs/2005.03085; Tom Britton, Frank Ball, and Pieter Trapman, "A Mathematical Model Reveals the Influence of Population Heterogeneity on Herd Immunity to SARS-CoV-2," *Science* 369, no. 6505 (August 14 2020): 846–49, https://science.sciencemag.org/content/369/6505/846.long.

67. Haley E. Randolph and Luis B Barreiro, "Herd Immunity: Understanding COVID-19," *Immunity* 52, no. 5 (May 19, 2020): 737–41, https://doi.org/10.1016/j.immuni.2020.04.012.

68. Gomes et al., "Individual Variation in Susceptibility or Exposure to SARS-CoV-2."

69. Mandavilli, "What If 'Herd Immunity' Is Closer Than Scientists Thought?"

70. "Coronavirus Disease (COVID-19): Serology Q&A," World Health Organization, June 9, 2020, https://web.archive.org/web/20201101161006/https://www.who.int/news-room/q-a-detail/coronavirus-disease-covid-19-serology.

71. "Coronavirus Disease (COVID-19): Herd Immunity, Lockdowns and COVID-19," World Health Organization, updated October 15 2020, https://web.archive.org/web/20201223100930/https://www.who.int/emergencies/diseases/novel-coronavirus-2019/question-and-answers-hub/q-a-detail/herd-immunity-lockdowns-and-covid-19.

72. "Great Barrington Declaration."

73. World Economic Forum, "Common Trust Network," December 20, 2020, https://www.weforum.org/platforms/covid-action-platform/projects/commonpass.

74. Rockefeller Foundation, "National COVID-19 Testing Action Plan—Strategic Steps to Reopen Our Workplaces and Our Communities," April 21, 2020 (PDF), https://www.rockefellerfoundation.org/wp-content/uploads/2020/04/TheRockefellerFoundation_WhitePaper_Covid19_4_22_2020.pdf.

75. Jillian Kramer, "COVID-19 Vaccines Could Become Mandatory. Here's How It Might Work," *National Geographic*, August 19, 2020, https://www.nationalgeographic.com/science/2020/08/how-coronavirus-covid-vaccine-mandate-would-actually-work-cvd/.

Chapter Nine: Take Back Control

1. Van Hoof, "Lockdown Is the World's Biggest Psychological Experiment."

2. Arjun Walia, "Edward Snowden Says Governments Are Using COVID-19 to 'Monitor Us Like Never Before,'" Collective Evolution, April 15, 2020, https://www.collective-evolution.com/2020/04/15/edward-snowden-says-governments-are-using-covid-19-to-monitor-us-like-never-before/.

3. Carl Zimmer and Benedict Carey, "The U.K. Coronavirus Variant: What We Know," *New York Times*, December 21, 2020 (archived), https://archive.is/dMEdJ; Apoorva Mandavilli, "The Coronavirus Is Mutating. What Does That Mean for Us?" *New York Times*, December 20, 2020 (archived), https://archive.is/4zjFT.

4. Colin Fernandez, "'Show Us the Evidence': Scientists Call for Clarity on Claim That New Covid-19 Variant Strain Is 70% More Contagious," *Daily Mail*, December 21, 2020, https://www.dailymail.co.uk/news/article-9073765/Scientists-call-clarity-claim-new-Covid-19-variant-strain-70-contagious.html.

5. Matt Ridley, "Lockdowns May Actually Prevent a Natural Weakening of This Disease," *Telegraph*, December 22, 2020 (archived), https://archive.is/d9otf.

6. Americans for Tax Fairness, "American Billionaires Rake in Another $1 Trillion Since Beginning of Pandemic," Children's Health Defense, December 14, 2020, https://childrens healthdefense.org/defender/american-billionaires-another-1-trillion-since-pandemic/.

7. Nicholson Baker, "The Lab-Leak Hypothesis," *New York* magazine, January 4, 2021, https:// nymag.com/intelligencer/article/coronavirus-lab-escape-theory.html.

8. "More Than One-Third of US Coronavirus Deaths Are Linked to Nursing Homes," *New York Times*, updated January 12, 2021, https://www.nytimes.com/interactive/2020/us /coronavirus-nursing-homes.html.

9. Merritt, "SARS-CoV-2 and the Rise of Medical Technocracy"; Rancourt. "All-Cause Mortality During COVID-19"; Yanni Gu, "A Closer Look at US Deaths Due to COVID-19."

10. Arjun Walia, "Another Vatican Insider: COVID Is Being Used by 'Certain Forces' to Advance Their 'Evil Agenda,'" Collective Evolution, December 28, 2020, https://www .collective-evolution.com/2020/12/28/another-vatican-insider-covid-is-being-used-by -certain-forces-to-advance-their-evil-agenda/.

11. Dr. Jospeh Mercola, "Mind to Matter: How Your Brain Creates Material Reality," Mercola .com, January 17, 2021, https://articles.mercola.com/sites/articles/archive/2021/01/03 /dawson-church-eco-meditation.aspx.

12. Lucy Fisher and Chris Smyth, "GCHQ in Cyberwar on Anti-Vaccine Propaganda," *The Times*, November 9, 2020, https://www.thetimes.co.uk/article/gchq-in-cyberwar-on-anti -vaccine-propaganda-mcjgjhmb2; George Allison, "GCHQ Tackling Russian Anti-Vaccine Disinformation—Report," *U.K. Defence Journal*, November 10, 2020, https://ukdefence journal.org.uk/gchq-tackling-russian-anti-vaccine-disinformation-report/; Nicky Harly, "U.K. Wages Cyber War Against Anti-Vaccine Propaganda Spread by Hostile States," *National News*, November 9, 2020, https://www.thenationalnews.com/world/uk -wages-cyber-war-against-anti-vaccine-propaganda-spread-by-hostile-states-1.1108527.

13. David Klooz, *The COVID-19 Conundrum* (self-pub., 2020), 71.

14. "Over 200 Scientists & Doctors Call for Increased Vitamin D Use to Combat COVID-19," VitaminD4all.com, December 7, 2020, https://vitamind4all.org/letter.html.

15. "Over 200 Scientists & Doctors Call for Increased Vitamin D Use."

Index

About the Authors

Dr. Joseph Mercola is the founder of Mercola.com. A family physician, bestselling author, and recipient of multiple awards in the field of natural health, as well as primary author of the recently published peer-reviewed paper, "Evidence Regarding Vitamin D and Risk of COVID-19 and Its Severity," Dr. Mercola's vision is to change the modern health paradigm by providing people with a valuable resource to help them take control of their health.

Ronnie Cummins is founder and director of the Organic Consumers Association (OCA), a nonprofit, US-based network of more than two million consumers dedicated to safeguarding organic standards and promoting a healthy, just, and regenerative system of food, farming, and commerce. Cummins also serves on the steering committee of Regeneration International and OCA's Mexican affiliate, Vía Orgánica.